Grow It!

Also by Richard W. Langer

THE AFTER-DINNER GARDENING BOOK

THE
BEGINNER'S COMPLETE
IN-HARMONY-WITH-NATURE
SMALL FARM GUIDE—FROM
VEGETABLE AND GRAIN GROWING
TO LIVESTOCK CARE

Grow It!

RICHARD W. LANGER

Illustrations by Susan McNeill

AGRICULTURAL CONSULTANT:
Wayne Stayton of Wayjan Farms

GALAHAD BOOKS   NEW YORK CITY

Åt Tant och Farbror Larson
Som lärde mig uppskatta livet på landet.

Contents

Introduction

For, eschewing books and tasks,
Nature answers all he asks;
Hand in hand with her he walks,
Face to Face with her he talks,
Part and parcel of her joy,—
Blessings on the barefoot boy!
—JOHN GREENLEAF WHITTIER

PERHAPS NOT SINCE THE FALL of Babylon have so many city dwellers wanted to "return" to the country without ever having been there in the first place. For the first time, the new generation reverses youth's traditional flow toward the city in search of opportunity. A cry of "back to the soil, to real life" leads the exodus. But just as the mechanically unskilled peasant floundered when tossed into the technocratic mechanism of the city, so today's new urban peasant, unskilled in agrarian survival, flounders when released in the meadow. What is lacking is a roadmap, a handbook for survival on the farm. Where once the craft of working the land and reaping its harvest was passed from father to son, today the torch of knowledge and experience consists of how to get credit cards and fill out income tax forms.

So how do you manage to jump off the technocratic trampoline and land in the country still standing on your feet? Returning to the land, homesteading, in the true sense of the word, be it on five acres or five hundred, getting to know mother nature again—these are things both

age-old and at the same time new as never before, because we have been
away so long, more than generations, some of us.

How does one begin? And where? Does it make a lot of difference
what part of the country you choose? How much land do you need for a
row of beans? How many plants for a blueberry patch? A field of sunflowers for you and your chickens? And how do you care for a baby
chick or a hive of honeybees? How on earth do you milk a goat? How
do you plow and fertilize and harrow the fields, working them to yield
good grain and fodder?

Do you have to do all that, anyway, to live off the land? Well,
country living isn't all just stretching out in a sun-warmed grassy
meadow and taking it easy—not if you're going to do right by mother
nature. It's give and take, the natural cycle of life, and we're just beginning to learn that maybe we've done too much taking and not enough
giving. But homesteading is clean living, good, earthy, sun-drenched
work, leaving you with that happy kind of tiredness at the end of the
day. And it's good play, too, reaping the whole cornucopia of rewards
that nature gives back to those who work with her. That's true if you're
homesteading a simple weekend family farm of a couple of acres—just
about right for a large vegetable plot, a berry patch, an orchard of half
a dozen trees or so, maybe a beehive or two, and a small flock of Crested
White ducks that don't need daily tending. It's equally true for a spread
of a hundred acres you've still, in spite of the way the land is running
out in this world, managed to find somewhere and hope to build up
alone or work communally into a better place for people to live.

The pages that follow are written as a kind of "boy scout manual" for those planning to return to the land. The ideas are for dream-
ing and doing. You may want goats and chickens and pigs and bees, or
none of these at all, but a big field of sweet corn instead. You may want
a pond or a wood or a grape arbor, or all three. Whatever your dream,
plan your escape to the country now, even if you can't move out yet.
Planting an orchard, building a pond and livestock shelters, getting the
fields ready—these are all projects that take time, but can be worked at
during the weekends. Your fruit trees, for one, will take several years to
bear. If you can plant them during vacations from the city, they may
well be waving their blossoms in greeting when you come home to stay.
The rhubarb patch can be started one spring and its harvest toted back
to the apartment for years to come. So can a summer's tomato crop, the
yield from your strawberry patch, and pumpkins from among the corn.

And if you're already settled in on your farm, here are ideas for bettering your homestead—nature's way. What a field of clover can do for your land, for instance, and how much grain for cereal and bread you'll get from a half acre. What a milch goat will need to supply your milk pail happily, or honeybees to stock your larder of sweets.

Although there is no virgin country left to pioneer, we can still reclaim land worn out from man's wasteful rapacity and replenish nature's store of goodness. And there's no need to give up the good things technology has offered us—there are many. There is only the need to temper the uncontrolled progress that today threatens to multiply our waste and destruction to the point of no return, to a point where nature will stop fighting back.

Nature is flexible. If man wants paradise, she will oblige. If he wants despair and destruction, she will oblige. Neither will come overnight. Nevertheless, change marches on steadily, unrelenting. It is still possible to decide which road will be taken, but not for long.

Of course you alone cannot save the world. But you can do your part. Grow good rye and wheat for your bread and noodles, grow good corn and oats for your livestock. Live and breathe, enjoy your life as you plow the rich earth, as you watch the crops rise from the earth to yield their bounty, as you celebrate Thanksgiving from a table groaning under the weight of your own fresh produce rather than the months' old, half-decayed cellophane varieties that the city dweller has learned to accept as fresh. And save the good seed from your rich yield to plant again, and harvest again, and plant . . . in the generous, timeless cycle of mother earth. Plant a tree where trees have been felled and only barren-seeming earth remains, waiting. Plant a bramble berry patch to harvest and share with the small wild things that have so little place left to call their home. Dig a pond and stock it with fish and frogs, and furry creatures will come to it thankfully. Sow squash for your own winter larder, and dry a few gourds to hollow for errant birds in search of a simple guesthouse in their restless flight from the encroaching macadam world. Raise a goat and give it a good home, not just a makeshift one somewhere. Raise a goose or two and some chickens. They will all do their part in return, enriching your land with natural fertilizer and giving you the simple necessities of life—milk, cheese, meat, warm down for your winter blanketing.

And then enjoy the honey pot your bees gave you as if you were Pooh. Enjoy a lazy day fishing the pond or tramping the fields and meadows. Enjoy the fruits of your labors. The world will be a better place for them. Today one small, well-kept farm, even of only a couple of acres, is worth a thousand cars spilling off the General Motors assembly line.

Apprentice yourself to nature. Not a day will pass without her opening a new and wondrous world of experience to learn from and enjoy. If we really mean to return to the land, let us not return empty-handed. Let us return to it something of value, some token of our alliance with mother earth. What has been termed organic farming is precisely such a recognition, although perhaps a misnomer. For organic farming, its name notwithstanding, uses both organic and inorganic material. All plant and animal life does. The purpose of the so-called organic method is simply to work hand in hand with nature, to avoid using destructive chemicals. More properly perhaps it should be called natural farming. But the name of the game isn't semantics, it's the healthier, saner, more rewarding life that comes of making your own way in the world and knowing that you've done a day's work that counts. So bring a small house gift when you come to the land—two hands full of willingness to do right by the land, the hands of nature's apprentice.

Grow It!

The Land

THE TIME IS COMING when farming will be done in the city. There's already talk of multistoried urban farms. These skyscraper produce factories will raise swine on one floor, chickens on another, with strawberry plants in between, all in one fully integrated, coordinated, complementary, scientific, and unnatural pseudo life cycle. The amount of land needed will, of course, be less. The quality of food produced . . . well, let's leave that to your imagination.

The people operating these juggernauts of scientific agriculture will know very little about nature, and even less about land. After all, why should they? It's not even inconceivable that in some distant civilization knowledge of plants growing in anything but test tubes or animals living in anything but concrete boxes will have vanished from man's experience—if he survives that long.

Meanwhile, back on the farm, to understand the land is to be able to work it well, to build it up, to give nature a chance. Few men today will ever be able to say, "I left the land in better shape than I found it." You will be able to. And a greater deed can no man accomplish.

SOIL

Take a spade and go for a walk around your fields. Stop at random and make like a gopher. Dig a hole a foot or so deep. Pick up a good handful of soil from the top of the hole and one from the bottom. Crumble it in your hands, smell it, check the color. This is mother earth, your share of it to care for, to feed, to harvest. Get to know the soil. Your farm will be the richer for it.

Soil consists of both organic and inorganic material. The two are complementary and essential for life. The organic, however, developed from the inorganic over the millions of years the earth has spun in its orbit. To give a simplistic description, all the destructive forces of nature—earthquakes, floods, frost heaving, wind and water leaching, glaciers, torrid heat, you name it—converged upon the rocks, which over the span of millenniums crumbled. The most solid of rocks has microscopic pores for moisture to seep into. Freezing water expands, rupturing the pores. Even where water doesn't freeze, it expands and contracts with thermal changes. A coffee-cupful of water shattered a house-sized boulder here, a thimbleful of water a six-foot rock there, not into a million pieces at once, but a million pieces in a million years. Glaciers ground the boulders into the earth, pulverizing pebbles between them as the millstone does wheat. From this stone flour came soil.

Somewhere, sometime, amino acids and enzymes entered the picture, catalyzed by the energy of lightning from a cauldron of nitrogen, oxygen, and carbon. For aeons enzymes burst forth and died during the crashing of storms, able to survive but the briefest of time, until some of them developed the facility to reproduce. And reproduce they did, until again sometime, somewhere, bacteria and viruses developed. The bacteria could draw energy from both the minerals in the crumbling rocks and the atmosphere.

Specialization set in and bacteria became more complex. The most primitive plants evolved, some of them perhaps nothing more than jumbo-sized bacteria. Eventually came lichen and mosses.

The die was cast. Mother earth turned in a direction that was to dominate her course from that time to the present. For the primitive plants died, leaving organic matter to mix with the inorganic. Life-bearing soil formed, slowly, in minute quantities, giving more complex plants a chance to evolve. Perhaps had the crystalline viruses been more efficient than the bacteria, they would have begun to dominate the world at that crucial juncture. Perhaps those first primitive plants would never have evolved. Perhaps instead the viruses would have increased in complexity. Perhaps the whole chain of life would have developed along inorganic lines instead of as we know it. Perhaps a super-complex ruby would be farming your field today, leading a plow drawn by an emerald to prepare the earth so it could grow rock crystals. Perhaps if man does not learn to recognize the value of life and nature, mother earth will yet have to start over. Maybe it will be crystals next time.

What happened this time, however, was that the roots from primitive plants began to work on the rock pile. A growing root, although it obviously prefers soil, can split a rock if it has to in its search for nutriments. This process, combined with the action of numerous acids released in vegetative decay, helped raise even further the organic content of what was slowly becoming soil. As it did so, more complex plants could develop and in turn be interred in the soil. The hills of mother earth were greening.

If you're serious about returning to the land, it will mean trying to help counteract the destruction spawned by today's total technocratic *zeitgeist*. You must help nature keep her mantle of green. Without it there is no future.

Good soil is almost one-fourth air. This, you'll agree after tilling a few acres, doesn't make it as light as a soufflé. But the air is essential to plant growth—yet another reason why organic matter in the soil is so vital. Not only does it supply nutrients, it facilitates aeration. And without aeration bacteria and fungi could not break down the organic matter itself. Aeration also aids in oxidation, one way the minerals in the inorganic matter are freed for plant use.

Basically then, the more organic material in your soil, the more plant life it can support, which in turn contributes more organic material so the soil can support even more plant life. . . . There are, of course, limits to all this, but the more you grow, the more you can grow, and the better quality it will be—as long as you keep it natural.

Tilth. The physical quality of your soil, the tilth, is what you're looking at when you're digging around on your land to see what's what. A soil in good tilth will be full of decaying organic matter. It crumbles easily in your hand, breaking down into pea-sized particles. You can crumble these particles even further with a little finger work, for each particle is many small granules of sand. And since the particles do not fit together smoothly, like cement blocks, but loosely, like a jumble of macaroni, there are empty spaces to hold water and air.

Good tilth gives your soil plenty of aeration and water-holding capacity. The greater the degree of aggregation and number of ameliorated soil particles, the better the soil tilth. And the better the tilth, the better your crops.

Soil Type. Rub the soil on your farm between your fingers. The gritty feel is the sand. When the soil is dry, the slightly slippery-feeling talcum-powder-like component is fine silt. When the soil is wet, the plastic, flexible quality is produced by the clay.

The study of soils can fill a book by itself, it has already filled many. For the apprentice farmer, however, the basic threefold classification of sandy, clayey, and loamy soil suffices. What you want, ideally, is loamy soil.

SANDY SOIL. Approximately 60 percent sand, 20 percent silt, and 20 percent clay. Drains too fast.

CLAY SOIL. Approximately 60 percent clay, 20 percent silt, and 20 percent sand. Doesn't drain enough.

LOAM SOIL. About 40 percent silt, 40 percent sand, and 20 percent clay. Drains at just the right rate to keep the soil moist without its being sopping.

You can't change the actual proportions of your soil too greatly. Unless you're going to truck in a couple of thousand tons, say, of silt. But with careful tillage—fertilizing well with manure, both green and barnyard, composting, liming, and using crop rotation—vast improvements can be made in almost any soil.

WATERSHED

Water is the very life of any farm's existence. Without it you have a desert. After all, a desert is potentially rich farmland, its mineral content usually being very favorable. Irrigation has been used in the Southwest of the United States and in many areas throughout the world to make arid and semiarid lands bloom. This form of intensive agriculture can be very productive. It is also extremely expensive, however, and therefore not a viable alternative for the small homesteader. But reclaiming dehydrated soil is only one side of the ecological coin. It's the other side, preventing the destruction of the watershed in the first place and the subsequent denuding of the land, that is your primary concern.

A large part of the Sahara was once the grain bowl of the Roman

Empire. Poor farming and overgrazing set in motion the erosion cycle that culminated in today's barren desert.

Our own Midwest suffered dire consequences when the Great Plains were plowed up and the native grasses that had held and developed the soil for centuries were destroyed. Some of these grasses had root systems so extensive that if all the hair roots of one single stalk were laid end to end, they would reach over three hundred miles. This means—to play the traditional game of pictorializing incredible numbers—the roots from the Plains grass growing on just a few square yards of soil could be wrapped around the equator several times. But the roots weren't stretched end to end. They were woven together in clumps. They were the warp and woof of the land.

Along came the plowman, planting domesticated grain grasses and such short-rooted crops as corn. Between crops the soil was left exposed to the destructive forces of sun and wind. In New England, where American farming began, this might have been no great catastrophe. Rainfall was heavy enough so that it would have taken centuries for the land to be destroyed. In the more arid Plains states the consequences were severe. Within only a few decades the world's greatest grasslands had become a dust bowl.

You too can destroy land, in a decade. Or, working hand in hand with nature, you can turn your farm into a small garden of Eden of ever-increasing abundance.

Water for your fields is dependent on the watershed. The watershed is a drainage basin or series of such basins defined by the topography of the land. Rain supplies the water for such a drainage basin.

Given two parallel rows of hills, for instance, the rain will flow off the hills into the valley between them. At the bottom of the valley, if the surface area of the hills is large enough, the runoff will form a stream. If the runoff is small, there might be only a swamp, or just plenty of subsurface water.

The water from this basin will drain out of the valley, either in the form of a brook or as subsoil seepage, into a larger watershed. Streams flow into creeks that flow into small rivers, and then into larger ones that finally find their way to the ocean.

Evaporation from the ocean, the rivers, and even the soil and the plants of your small farm rises into the atmosphere, which, when it reaches saturation, lets the water fall back to the earth in the form of rain. And the rain runs off again from the hills into your watershed. It's all part of one grand natural cycle.

Your farm as a mini-watershed is essential to the continued existence of this cycle. Even as great a watershed as the one for the Missouri
River, covering roughly 500,000 square miles, would eventually destroy
itself if it were surrounded by poor drainage and erosion.

The piece of land you buy, no matter how small, comes complete
with a watershed, or a part of one, even if you've settled in a wide, open
place. Your job is to find out what that watershed is like, then, in almost every case, to improve it. Watershed management involves work
you might not have considered part of country living. But it also af-
fords an immense amount of rewarding pleasure. Part of many a farm's
water program is the pond, an invaluable source of fish and plants such
as watercress, but also a place for ice skating in winter and a cooling
summer dip on the way home from the fields.

You can't be expected to control underlying geological formations,
yearly precipitation, and other natural regional variations. But there are
four basic characteristics of a good watershed that you can regulate
through careful planning. These are also the characteristics to watch for
when first checking out your land.

Quality of the Runoff.   Land that has been carefully planted and
well managed will have a runoff that is clear. Even after a rainstorm
surface water draining into a creek or pond will be clean and pure. It
may be clouded with soil particles immediately following a particularly
heavy burst of rain, but in that case it should settle out within an hour
when a glass or jar full of it is left sitting still. Muddy runoff comes
from poorly planted, sick soil. It will silt up your streams and ponds,
kill your fish, take away your topsoil, and generally deteriorate your
farm. Not a kind word can be said for allowing this kind of poor
drainage.

Noncyclical Stream Flow.   The greatest changes in nature are
wrought by cycles. The alternate freezing and melting of water in the
minute cracks of boulders will eventually crumble even granite. Alternating high-flow runoff and dry periods will in the same way wash away
the best soil. The flow of a stream through a watershed should be fairly
regular year round. Of course, it will peak in the spring if you're in
snow country, and drop to some extent toward the end of summer. But
a good watershed should have enough organic matter to absorb any
threatened water accumulation. When the snows melt, there should be
no torrent of water rushing from the land surface into the stream.

Rather the melting snow should be absorbed by the soil as it warms slowly to water and then gradually make its way beneath the ground to seep out into the stream.

Soil Condition. The soil carries water throughout your land the way the vessels of the body carry blood throughout you, except instead of being pumped around by a heart, the movement of water is produced by a cycle of gravity and evaporation. But just as a limb cut off from its blood supply becomes gangrenous, so land cut off from water dies. Even layers of hard clay, for example, are not incapable of supporting plants, if such soil is revitalized by the introduction of organic material capable of water retention. Clover sod plowed under, followed by rye and buckwheat or wheat plowed under alternately with generous liming—one of the most effective combinations of crops for this purpose—can give miraculous results in a couple of years.

As mere clay, soil will not absorb water. Rain and melting snow run off quickly, taking even more soil with them, and things just become steadily worse. But after even one year of soil revitalizing, enough organic material will be plowed under to act as a sponge nucleus, absorbing rainwater for use by new plants, whose fine root system in turn holds the soil in place and works as a sieve, catching minute soil particles in the water. Muddy runoffs, brown creeks, and sludge-filled ponds become clear, and life begins to abound.

Vegetation. The best natural protection for a watershed is grasses, legumes, and trees. All three permit utilization of the watershed for other crops as well. Properly managed grassland and meadows afford nutritious grazing grounds for your livestock. Legumes not only make hay and when plowed under build soil, but those like sweet and white clover are also excellent hunting grounds for your honeybees. (Honeybees can't gather the nectar from red clover because their tongues aren't long enough.) Trees yield lumber when felled selectively so as not to upset the watershed balance. A hillside orchard combines both good watershed management and a yield of fruit and nuts for your table.

Keeping vegetation balanced on your acreage is the obvious sign of good management. Overgrazing a pasture one year can undo ten years of soil-building. For optimum utilization of the land, contact your local Soil Conservation Service. Conditions vary from area to area even within a county, but these people know your land and will be able to suggest the specific tack to take for the prevailing conditions, as no general manual on farming could hope to do.

THE POND

The big country kitchen has always been the city dweller's dream of the farmhouse family room, and rightly so. But there is a second place where all the farm inhabitants, human and animal, gather—the pond. It's the old watering hole to the livestock; to you it's for fishing, raiding nature's ice bucket, gathering watercress, fire-fighting if need be, not to mention swimming in summer and ice skating in wintertime. Besides all this, a pond is often a vital link in maintaining a well-balanced watershed.

Many farms come with an old pond, natural or man-made. But traditional farm ponds, part of the rural fire-fighting equipment and the water troughs' supply system, are located close to the barn. Often the barn is halfway up a hill, with the pond below it. Runoff from the barn area may make the water unsuitable for swimming. A second pond never hurts. And, of course, your farm may be one of those that never had one in the first place.

Contact your local Soil Conservation Service or Agricultural Extension Service and discuss pond-building with them. They can help you decide on the best location, size, and layout of the pond. Check up on any government assistance that might be currently available to help you pay for pond development. Since in many cases ponds are vital in flood control as well as for general land improvement, the government will often offset part of your expenditures. But always remember to check these deals for attached strings. Some government fish-stocking plans, for instance, call for your ponds to be open to the public as recreation areas during parts of the year. That may or may not be fine with you. The point is, make sure you know what you're getting into.

A pond can be built at the head of a spring, or in an easily dammed gully with a small, trickling stream—but don't try to dam off a stream of any size. It can't be done without concrete and good engineering. The important thing for a dam is that the soil be of a variety that holds water well and that the core of the dam be packed solid.

Even if there's no direct source of running water on your property, you might be able to build a "sky pond." A sky pond is one fed solely by runoff. A ten- to twenty-acre watershed and medium rainfall should be able to supply a ten-foot-deep one-acre sky pond. However, it may take a bit of sophisticated soil management. If your hills are full of gullies or in a state of progressive erosion, you'll have to remedy that first. Muddy runoff from overcultivated fields or bare, steep ones will fill a sky pond with silt in a few years. Mud flats are of no benefit to your farm.

Whatever the means of feeding your pond, remember it can't be just a shallow puddle. A deep one-acre pond (just about the minimum practical size) is far superior to a two-acre wading pool. The latter may look nice enough, but leaving out for a moment the fact that you can't swim in it, a shallow pond has a high evaporation rate and is a fabulous mosquito breeder. A deep pond, on the other hand, will support plenty of fish to eliminate this winged problem while supplying your table with some interesting variety. In a good one-acre pond you can catch two hundred to two hundred and fifty pounds of fish a year, and the yield increases progressively with larger ponds as long as they're deep enough so as not to freeze solid in winter. A fish crop can be harvested as regularly as anything else on the farm. In fact, to keep the pond well balanced, fish you must! Sitting in the shade of the old weeping willow with a bamboo pole dipping its line lazily in the pond is a "job" that needs to be done—it's pretty hard to imagine another essential farm task as pleasant.

There are other reasons besides the comfort of the fish population for making a pond deep. For one thing, the ratio of surface area to total pond volume makes a big difference in its watershed value, particularly in a sky pond. A flat, pancake-like pond loses immense volumes of water through evaporation, so much in fact that in summer it's apt to turn into a swamp. Of course, you don't want a pond like a swimming pool, with perpendicular sides, either. Sun-warmed shallows make the best breeding ground for many fish. What is important is that the center be deep. This gives the fish a place to go with enough oxygen for them in the wintertime when the pond freezes over.

Pond life in general needs clear water, not silt. This means land leading down to the banks should be planted to a good deep-rooted cover crop like clover that will assure a runoff as clear as the rain itself, with only a few nutrients added through ground seepage.

Berry bushes and trees, on the higher banks where their roots will not be drowned, invite the homesteading of wildlife, from birds to raccoons and rabbits. Try to make one side of your pond into a dense multilevel jungle by planting low shrubs, such as hazelnuts, and willows toward the banks, wild cherry, crabapple, and other fruit and nut trees farther back, and a good stand of hardwood and evergreens to give more distant cover.

All this is a lovely sketch on paper. But what about actually digging a pond? Let's say you're starting from scratch. You've arranged for a bulldozer operator to come in and do the excavating. It will take him only a day or two to scoop out a small pond.

The earth removed from the bottom will be piled up at one end as a dam. The slope of the soil on the water side should be considerable, perhaps as much as 60 degrees. The side away from the pond can be steeper. Get specific information for your site from the Soil Conservation Service. It's a good idea to flatten the top into a roadway rather than leaving a mound. Then too, provision must be made for extra water to leave the pond. It should never be allowed to run over the dam. An overflow culvert, or series of small concrete pipes buried under the roadway at the top of the dam, will serve as a good spillway. The spillway should be at a level such that the water overflows with some regularity. You don't want the high-water mark to be so high that the pond overflows for only a day or two with the spring floods. A slow, small, steady trickle is what you want for the best overall pond balance.

With this steady flow of water through the culvert, you will have a natural spot for a watercress field on the other side of the dam from the pond. Good watercress needs cool, clean, continuously flowing water. Water from a limestone area is best. However, throwing crushed limestone into a stream upwater from the cress will give good results too. Even if you don't build a pond right away, you can still have a cress bed, wherever there is clear running water.

Watercress is an easy no-muss no-fuss farm crop. Plant it once and you'll probably be able to harvest for decades. Your best bet is to get established sprigs and plant them at the edge of your stream. However, it's not much more difficult to start them from seed. Stick a couple of pots of soil into your stream so they are just slightly higher than the water level. Sow the seeds. Once seedlings develop, transplant them to the banks of the stream. You can sow seeds directly into the banks— but in that case be prepared to do most of your harvesting about half a mile downstream. Plants mature in fifty to sixty days.

While you're waiting for the water to rise enough to overflow so you can grow your watercress, fertilize your dam well with manure and sow to rye and clover or crown vetch to prevent erosion. The second year, once the cover crop is well rooted, plant a row of fast-growing willows.

It will take your pond anywhere from a month to half a year to fill with water, depending on the watershed conditions for your particular site. As shallow water begins to fill the pond, you'll notice life joining it. First insects like water skaters and tailors; then frogs and toads appear, seemingly from nowhere, to feed on the insects. Word of a new place to pioneer somehow spreads to neighboring ponds or swamps. Algae grow in the sunny shallows. Birds come to splash and bathe; if you're lucky some of the rapidly vanishing long-legged waterfowl will stop by for a frog feed. In their droppings and under their "toenails" they carry seeds and semimicroscopic animal life, such as amoebas and hydras, from other ponds, stocking yours with innumerable new inhabitants.

The pond will rise and fall as the water settles in. Much of the new life will die on the drying banks, only to be swept up in the water as it rises again, this time to feed the remaining life. Winter will come, and as the ice covers the pond, the frogs will vanish, the insects disappear, and plant life for the most part will die down. Even if you are in an area of very little freezing weather, you'll see the yearly cycle of the pond stilled into semihibernation.

But come spring, new life awakens the pond. Vegetation will be much denser than the first year. Colorful dragonflies and other flying water insects join the pond community. The frog population explodes into the thousands. Chances are you'll even find some minute fish in your pond by the time midsummer rolls around. All without your ever having put a single animal into the water. But you'll want to introduce some varieties of your own to speed things up.

Now that plant and animal life are established, you can plan on stocking your pond, since the fish will have something to eat besides each other. If the water is deep enough, trout will take; if not, bass and sunfish certainly will. For the best selection, have your pond stocked from a local hatchery. If you can't do that, go fishing in another pond for the day and stock yours with the catch. A dozen bass will do a good job of populating a pond in a couple of years.

Besides stocking with fish, you will also want to fertilize your pond. Fertilize a pond? Yes. Like any other "field" you expect to harvest, your fish pond needs fertilization. It's not like sprinkling fish food into an aquarium. What you want to do is give nature a boost so the pond can support more and bigger fish on its own. Fish-pond fertilizer is the same as that for your land. Manure, compost, and ground rock phosphate will help nourish strong plankton, which in turn feed the fish. The exact amount of fertilizer to use varies greatly with pond size and layout. General rules are pretty useless, except to say that a monthly summer feeding is better than dumping everything in at once. If you can, check with whoever stocked your pond for advice on fertilizing it. In any case, remember you can overfertilize, making the water dark and scummy. This not only deprives the fish of their oxygen, but makes for lousy swimming.

Once the natural cycle of a pond is well established, your only task, except for the occasional fertilizing, will be fishing. Regular harvesting of the pond makes for bigger and better fish—not to mention tastier ones.

WOODLAND

On almost every farm there's a small stand of woodland trees somewhere—in a gully area or on a knoll for some reason not suited to crop production. At least not crops in the usual sense of the word. But woodlands do provide timber for building or fence posts, and fuel for that crackling winter fire whose leftovers, the wood ashes, in turn make a very valuable potash-rich fertilizer for your vegetable garden. In the Northeast these woodlands may also supply maple syrup. Elsewhere they act as windbreaks, help in erosion control, and shelter local wildlife.

If there is a woods on your farm, build it up with good management. If there is none, start one on a plot you don't need for crops or pastures. A particularly good location from the standpoint of encouraging wildlife is flanking one side of your pond. Wild pheasant and other game birds may visit your forest park, and you'll see hare scampering and inquisitive squirrels on your fishing trips.

WOODLAND RESUSCITATION

Probably less effort is entailed in woodland management than that of any other "crop." Moreover, the tasks involved can be done almost any time, when you're not busy with something else and are looking for a good outdoors project for a sunny day.

Assuming you already have an acre or two of trees remotely resembling a woodland, the thing to do is improve the glade. Specific recommendations would vary from area to area, but the basic principles are to thin out the stand, permitting fewer but straighter trees to grow, and to remove all underbrush except from about a fourth of the area, left untouched for the wildlife. Eventually you may want to offer them a higher standard of living by planting more fruit trees and berry bushes; in that case, your wildlife shelter is a temporary, makeshift one for the couple of years it's going to take you to rehabilitate the woodland. Meanwhile, on a cleared quarter of an acre you can be planting new, more productive undergrowth such as hazelnut bushes, wild roses, and dense bramble berries for good nibbling. Hawthorn, wild plums, and persimmons are also excellent. Flowering dogwood, bayberries, and honeysuckle will not only attract wildlife, but will also enhance your own explorations of the woods, and maybe start you making scented homemade candles or fragrant potpourris for your country kitchen.

You want the woods to support wildlife. You also want to exclude livestock. Don't make the forest glade part of the pasture. A pasture should have its own shade trees. Livestock roaming the woods will nibble away seedlings, injure the bark and roots of young trees, and more than offset the addition of their fertilizing manure to the soil by trampling and packing the ground. You can't very well till the forest to make up for the damage.

If you're trying to rehabilitate an old wood, probably the first thing you'll notice is that the most desirable trees—the hardwoods such as oak, ash, hickory, sugar maple, and black cherry—are scarce indeed. That's because they have been felled without thought for the future. Your job is to maximize the growth of any few remaining ones. Some softwood evergreens, larch and fir, for instance, which do not make particularly good timber or firewood, also should be allowed to remain, to give esthetic balance to the woods and because they offer a variety of food and shelter for the wildlife.

One thing to remember from the outset in tackling your renewal project is that the edges of your woods protect it. You don't want to open the glade so much that the soil is suddenly exposed to new sun and wind conditions. Leave a dense hedge of trees and bushes around the borders. It's comforting for wildlife too.

Secondly, plan your cutting so that for the most part the poor specimens, extremely crooked or split trees and decayed ones, are the ones you remove. An exception to this rule is in your wildlife refuge. It will want a few rotten hollow logs. If there's a dearth of them there, haul your best specimens from a cleared section over to the reserve.

To space the trees for maximum growth you will need to thin the woods. Leave young saplings, for the future, wherever they are more than five feet apart. Older trees should be spread out so that the crown of any one is not touching its neighbor. For maturing trees, this usually means a space of ten to twenty feet between them.

Give the trees a helpful pruning. Remove dead or broken branches, and those within reach that simply make the tree too dense. This isn't the kind of thing you do yearly. But every old, neglected mini-forest should have it done once. The project will no doubt take several years of spare time, so don't try to set yourself a two-week rehabilitation deadline.

All trimmings and cuttings should be removed from the forest floor. Leave drying brush around, and you're inviting a forest fire. If you have a means of chipping the wood, the chips can be spread on the forest floor along with a heavy dose of nitrogen-rich fertilizer, such as bone meal, tankage, or dried blood. They will provide a good, slow, fertilizing mulch. If you have nothing to chip with, break the branches into as small chunks as you can and build a long-term compost heap of them at the edge of the forest. After it has decayed enough it can be returned to the forest floor.

ESTABLISHING WOODLANDS

If there's no patch of woods on your farm, start one. The best location is the worst one—that is, the spot that isn't much good for anything else. Trees will grow almost anywhere except on very poorly drained land. A gully or a rocky field can often be converted into a wonderful forest. Even an acre is enough to lose yourself in for a day, watching the wildlife, so don't decide to pass up a woodland because of limited space. And it's going to take longer than any other project on the farm, so start as soon as you can.

The first thing to do is to check with the State Forest Department to see what species they recommend for your region and whether they have a reforestation program that will cover part or even all of the cost of your stock. Once you've picked the location and desirable species—always remembering to include some black walnut and other slow-growing, high-value timber trees that will soon become extinct without the help of small local plantings—order your stock for spring delivery. You'll want to plant as early in the year as possible. Guidelines for planting the larger orchard trees are given in the chapters on "Fruit" and "Nuts."

And you'll want some trees larger than seedlings. They cost more and are harder to get sometimes, but they will add variety to what might otherwise be a field of Christmas trees. Also, a few older nut and fruit trees interspersed in your forest will be able to sustain wildlife much sooner than trees started from young, short whips. Plant some black locust as well; it's fast-growing and makes excellent fence posts. Just make sure it's *black* locust.

Good spacing for your forest, using three-year-old seedlings or transplants, is a criss-cross pattern, six by six feet to begin with. Even leaving more room around the bigger transplants that are giving your forest its head start, you'll need over eleven hundred trees per acre (another reason for starting your forest early). It's going to take a while to plant a couple of acres. By way of comforting comparison, it would take nature unassisted many hundreds of years to reforest a bare field. And chances are if you didn't do it, nature would no longer be able to. The cost of a hundred spruce seedlings—what you'd pay for a couple of pizzas—is a small price for giving nature back her own.

As the forest grows the first year, high grass will sprout between the trees. Young pine trees in particular are a sort of living fire hazard. Cut the grass and, the first two years, haul it away to help minimize the risk of fire. The cut grass can be put to good use by composting and then returning the results to the trees as mulch. If you need hay, you can cultivate the area between the trees with timothy and clover the first couple of years. Whatever you choose to do, keep the ground cropped close.

As the trees get taller, you'll have to thin the planting. A grid six by six feet will not support maturing trees. In fact, after fifteen years the stand should have been thinned down to about two hundred and fifty trees per acre. The rest you can sell or use for fence posts, firewood, and furniture.

Fifteen years? That's an awfully long time, you say. Who knows where you'll be fifteen years from now? True. But even if you're not on the same homestead then, you will have left behind you a small forest, part of which, like the black walnut trees, will keep growing for another hundred and fifty years. A utilitarian heritage of beauty that, had you not planted it, would never have existed. How many people do you know who have given nature back even a small forest?

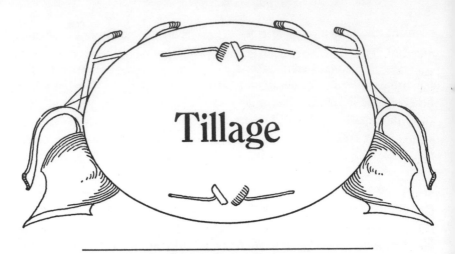

Tillage

He that by the Plough would thrive,
himself must either hold or drive.
—BENJAMIN FRANKLIN

ALL RIGHT, there you are with several acres of land in front of your nose, and maybe, by your side, a tractor and plow you decided were the proper things to have on a farm. What do you do? Plow, of course. But how and where? Well, first let's look at the equipment. Even if a neighbor comes in to help plow and seed the fields the first couple of years, you'll still need to know what the equipment is all about. A farmer is willing to help and more than willing to teach you, but he's not going to do the work while you sit back in your rocker on the porch. Also you'll find it's easier and less expensive than you think to get some of your own equipment. You'll certainly want to do your own plowing. After all, that's what farming's all about.

The simplest and cheapest tools to cultivate your land with are a spading fork or shovel and a rake. You dig up the soil and turn it over, trying to keep most of the topsoil, which is darker, richer, and looser than the subsoil, just where it was—on top. At the same time you are breaking up the clods of earth and loosening the soil to improve aeration. That's tillage, good old-fashioned, primitive style. With this help roots can grow more easily, and better roots mean better plants. In most cases, you'll find that the topsoil is a layer six to twelve inches thick. Very poor soil may have only a couple of inches worth. Some fortunate California valleys have topsoil thirty feet thick.

After the digging and clod-breaking, the ground should be loose and crumbly. This, however, isn't enough. You'll be planting tiny seeds or seedlings. The young early roots need more help than that. Rake the soil back and forth until all the particles to a two-inch depth are very small. The surface layer, especially in the vegetable garden, should have a particle size half that of a dime or less. However, don't tamp down the ground perfectly smooth or water penetration will be affected. Leave rake furrows on the surface to aid drainage.

All this is fine for a small vegetable plot. But think of it in terms of an acre, or even only half an acre, and just the thought will leave you panting. There is a solution to this backbreaking problem—albeit one that will shatter your peaceful country silence for the duration—a kind of small farmer's walking tractor or horseless plow.

EQUIPMENT

Rotary Tiller. The rotary tiller is a noisy but efficient little back-saver. This machine is almost a must for a garden of any size. It is simple to operate and easy to maintain. Usually it is no more than a power lawn mower, except that instead of cutting grass, it digs the soil for you. It will also dig in manure, both green and barnyard, compost materials, lime, and what have you.

Rotary tillers usually come with wheels mounted in front to pull the digging blades, or wheels in back to keep the machine steady. The front-puller models are easier and smoother to operate, in fact no more difficult to manage than a power mower. Those with rear-mounted wheels are less expensive. For a real back-jolting thrill, try one of the cheaper, older rotaries that don't use wheels at all, but drag themselves along by the digging blades—you balance them and steer the course. It's about like holding onto the tail of a rampaging drunk kangaroo.

You'll find a rotary tiller cuts your gardening work to a fraction of the time the shovel method takes; it still isn't sufficient when you're considering a field of, say, even five or six acres. What with waiting for the ground to be plowable and still trying to get the seeds in as early as possible to extend the growing season, you won't have the spare month or two you might need to rotary till half a dozen acres. You'll either have to trade labor with your neighbor or get a tractor and plow. The latter alternative is not as costly as you might think, simply because you don't need the latest super-rig. One of the old, semiabandoned

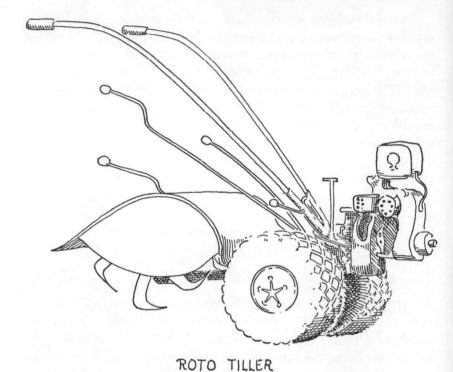

ROTO TILLER

tractors no longer economically viable for the professional farmer will do you fine. You'll even find that they are often cheaper than a new rotary tiller. A farm auction is a good place to pick up old equipment cheap.

Tractor. Tractors come in various horsepower, but this isn't really a good way of telling whether they will meet your needs. What you want is a tractor that can draw a one- or two-bottom plow, that is, a plow that turns one or two furrows at a time, respectively. It will do fine no matter what its horsepower.

Don't fool around with garden tractors or the oversized lawn-mower variety. They'll only be a waste of time and money. If you're going to get a tractor, go straight to the real thing.

There are several pieces of auxiliary equipment you might consider picking up to go with your tractor. Not all at once, of course. But country auctions are a lot of fun, and once you've been, you'll probably be hooked on them. Not just for equipment, but comfy old furniture for your home, and occasionally even livestock.

The first thing to remember in buying auxiliary equipment is that whatever you get has to be compatible with your tractor. If you have a small, older tractor capable of pulling a one-bottom plow, it's not going to do you any good to get a five-bottom plow unless you enjoy sitting with your tractor and plow in the middle of a hot field with your engine flat out, not going anywhere. Most equipment will have a tractor hookup compatible with your tractor. Note the word *most*—not all. Be sure to check that whatever you get can in fact be hooked up without major adjustments.

Plow. You may find your first major tractor attachment, the plow, comes with the tractor. This is because of the two major types, trailer plows and mounted plows; the latter, as its name implies, comes mounted on the tractor.

If they don't come as a combination, you'll probably have to get a trailer plow. Chances are you won't be able to find a mounted plow to fit. The trailer plow is preferred anyhow; it does a better job. On the other side of the coin, if you're tossing for it, a mounted plow is easier to maneuver.

Trailer plows are of two varieties, the moldboard and the disc. The disc plow is in Louis Bromfield's estimation just about the greatest step forward in tilling since primitive man first scratched furrows in the earth with a forked stick. It has several huge steel wheels five feet or so in diameter, all running on one axle, that literally chop up the soil, digging deep for aeration and breaking all the clods. It's particularly good for incorporating organic matter far down into the soil when trash-mulching, that is to say, burying field plants after harvesting by turning the soil over on top of them. Working like one of the old-fashioned spinach choppers of European kitchens, it prepares the ground for planting in one coordinated operation.

Unfortunately, as effective as they are, particularly on heavy clay soils, you'll see few disc plows around. Unless you're working with two, maybe three hundred acres or more, they're just too big and expensive. Think of them as thousand-acre rototillers.

This leaves you with the old traditional moldboard plow. There's nothing wrong with a moldboard plow, however, as long as you use it right and remember that plowing alone won't be enough to get your soil in shape for planting. You'll have to harrow the field as well.

The moldboard plow is the kind the urbanite expects to see on the farm. It has a blade or blades like half a heart, curved to dig into the

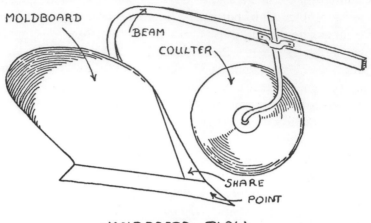

MOLDBOARD PLOW

soil, lift, and turn it over. At the lower edge of the "bottom" is the point on which it rides, and at the front of the moldboard is the share, which cuts the soil. You will no doubt break a few points while you're learning the ins and outs of using your plow. Cheer up. They're replaceable. Each bottom makes a furrow, casting a row of molded earth to one side. On the arm, or beam, holding the bottom and in front of it is a coulter. This is either a wheel or a mini-plowshare that precedes the plow to open the sod. The coulter doesn't always have to be lowered, but with heavy sod, or heavy mulch such as cornstalks, plowing without it is impossible. The wheel coulter is the preferred type since it won't snag cornstalks or long-stemmed green manure.

With a one-bottom plow you'll make one furrow at a time, with a two-bottom plow two at a time, and so on. Since a two-bottom plow will cut the number of passes you have to make up and down the field by 50 percent, it's well worth getting if you can. But again, don't buy a bigger plow than your tractor can handle.

Moldboard plows are divided into three general classes by the shape of the moldboard: stubble, sod, and general-purpose plows. The stubble plow has a short, steeply pitched blade that breaks the soil into relatively fine particles. However, it can only be used on land that has been cultivated for some time. Try to use it on sod or overgrown pasture land that has not been worked for decades, and its sharply curved moldboard will dig in rather than plow. You'll get nowhere. For the purpose of breaking sod, the sod plow has a very long, narrower moldboard with considerably less pitch. It won't pulverize the soil as well as the stubble plow, but it will turn over fields that could not be worked otherwise. Halfway between these is the general-purpose plow, the one you'll most likely use.

Chances are, since your neighbors will be working soil not too different from yours, if you get your plow at a local auction it will serve your purposes admirably. You may even find one lying in back of your barn, but in that case be prepared to do an awful lot of cleaning. You've heard of a greased pig. Well, how about a greased plow? A plow blade must be smooth and clean or the amount of energy needed to pull it through the field increases immensely. Always clean your moldboard after plowing. You don't have to dry it like fine china, but if it's going to stand around without being used for a while, it should be covered thoroughly with grease to prevent rust.

Your neighbor might very well have an old plow around he'll be more than happy to trade you for something. This is about the best way to acquire it, since you'll need him to show you how to plow. Don't be discouraged, it's a real art when done right, and will take quite a bit of practice. But then again, you wouldn't expect to be able to walk on your hands the first time you tried it either. And the switch from pushing a pencil in the city to working the soil in the country is certainly no less radical a flip-flopping of your world.

Harrow. If the plow is your supersized spading fork, the harrow is your streamlined rake. Harrows come in two main types, spring-tooth and disc. You may have to settle for a spring-tooth harrow in the beginning. But try to get a disc harrow if you can. Although its is both more expensive and more finicky about its maintenance, these drawbacks are more than compensated for by its versatility. Where two passes with a disc harrow might suffice for a good job on your field, using a spring-tooth harrow you'll have to go over the same field three or four times. And thick-stemmed crops such as corn or sunflowers can be stubble-mulched directly into the field again with the disc model after harvesting. That is to say, all the stalks and stubble are worked right into the soil by running the heavy disc harrow over the field a couple of times. The discs cut up the stalks and bury them. A spring-tooth harrow run over this kind of field would simply jam up.

The spring-tooth harrow looks like four huge combs with curved teeth, mounted one after the other, with a flexible backbone. As you pull

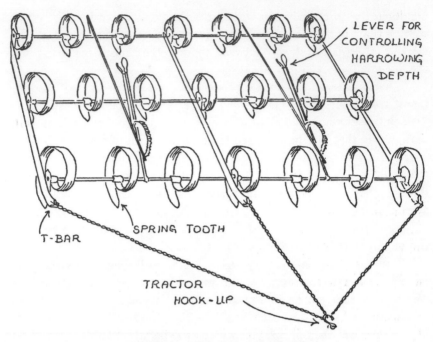

LEVER FOR
CONTROLLING
HARROWING
DEPTH

SPRING TOOTH

T-BAR

TRACTOR
HOOK-UP

SPRING TOOTH HARROW

the harrow across the field, these teeth break up the lumps of soil. Each tooth hits the soil, is pulled back, and snaps forward with a vibrating chomp, shattering the clump. The teeth all work together, a lot of busy fingers crumbling the soil. On a disc harrow the spring teeth have been replaced by discs that work the soil and break up the plow lumps.

Broadcaster and Grain Drill. You can sow most seeds by either broadcasting or drilling. Broadcasting is simply scattering the seeds evenly over the desired area. You fill up an old shoulder-strap bag and then walk the fields, scooping out a handful of seeds and flinging them from your hand in a wide arc. But the simplicity of this age-old method is deceptive. It takes quite a bit of practice, not only to sow the seeds evenly, but to scatter the right number of them per acre.

To help you out, there's a simple and inexpensive device that does your spacing for you, though you still have to do the walking. It's a bag with a crank and spout. You hang it over your shoulder and turn the crank as you stroll along; a continuous stream of seeds is forced out the spout in a fanlike pattern, to be buried by the elements. The instructions accompanying the device will tell you how fast to turn the crank and at what setting each size seed should be sown. The broadcaster is sufficient equipment to take care of several acres. If you have more than that to seed, you'll want a grain drill.

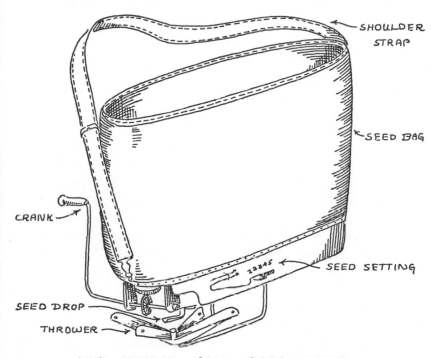

HAND CRANKED SEED BROADCASTER

A grain drill is also a reasonably simple machine and relatively inexpensive if bought used. Basically it's just a long horizontal hopper on wheels. Full of grain, it is pulled behind the tractor, and leaves the seed in a nice evenly distributed pattern known as drills. The seeds are buried by the action of melodiously tinkling chains dragging behind the grain drill. The drills here, incidentally, have nothing to do with holes, as one might imagine when thinking of a drilling machine; in this case the term refers to the long rows in which the seeds land. The grain drill has gears and openings that can both be adjusted to sow the desired number of bushels per acre. Usually you'll find the necessary information on proper settings engraved on a small instruction plate near the gears. When buying an old grain drill, be sure this plate is still on the machine, or get someone familiar with the model to show you what the different settings are.

If you get a grain drill, you'll also be able to use it to sow corn, sunflowers, and other row crops. In this case, however, you will have to stop up most of the holes, leaving open only enough to space out the rows properly. If you want rows forty-eight inches apart, for instance, then all the holes between those forty-eight inches apart will have to be blocked. There's another use the grain drill can be put to, with only one change of setting. You can convert it into a lime spreader, providing your model has a fertilizer hopper as well as a grain hopper. One with just a grain hopper doesn't have enough capacity to make it worthwhile.

Corn Planter. A grain drill specifically for row crops has been designed, called a corn planter. If you intend to plant several acres of, say, corn or sunflowers, and you can find a single- or double-row corn planter, it might be worth getting. But buy the grain drill first. It's more versatile, being useful for sowing your pasture and green manure crops as well as your grain, not to mention liming. And, as noted, it can be used for row crops with a bit of modification. A corn planter is adjusted similarly to the grain drill to drop its seed at so many bushels per acre in single rows from between a pair of inclined discs that furrow the soil and cover the seed.

Manure Spreader. If your herd of livestock ever increases to the point where you have fifteen to twenty-five large animals of the pig or goat size, a manure spreader will be worth your while. Old manure spreaders never die, they just sit out back and rust. Which is to your advantage. It might not have that sparkling color you see in the farm machinery catalogs, but an old rusted one in working condition won't cost you anywhere near as much either. In its simplest form a manure spreader is a wagon onto which you load the manure. As you pull it across the field, numerous steel fingers rotating on a shaft which takes its power from the tractor or from its own wheels fling the manure in an even pattern from the wagon's tailgate.

Cultivator. Crops that are broadcast, or drilled grain, will not need to be cultivated. However, row crops, with large spaces between the rows, will need some tending. Cultivation in the field is what in the vegetable garden is called hoeing. That is to say, the soil between the row crops is broken up to keep it loose and friable, or crumbly. At the same time, weeds are turned over and killed. The cultivator is mounted on the tractor and does the job mechanically. In appearance it's nothing but a series of small shovel-like blades strung together on a frame. A rotary hoe works on the same principle, wheel-mounted fingers doing the job.

Be prepared to cultivate part of your crop as well as the spaces between it the first few times around, because until you get the hang of it, you'll probably be digging up some of your crop plants along with the weeds, particularly at the ends of the field.

Mower. The mower is a giant version of electric barber clippers. Parallel sharp teeth slide back and forth, cutting anything that gets between them. A mower is attached to the side of the tractor and takes its power from the engine through a hookup. It cuts your hay.

Grain binders used to be available for this purpose. They both cut the grain and bind it ready for field-curing. However, with the advent of the combine, which cuts and threshes at the same time, grain binders have become scarcer than the old proverbial hen's tooth.

To cut grain you'll either have to use a cradle scythe, which will do fine on a couple of acres, or get a neighboring farmer to run his combine over your field. If you try to use your mower on grain, you'll find that the grain heads will all shatter, leaving a well-sown field, but no crop for flour.

WORKING THE FIELDS

Now obviously you won't get all your farm equipment at once. You may not bother to get it at all if you're trading work with your neighbor for help on your fields. But if it's work you're trading, you'd best have a little know-how, or it will do neither your fields nor your neighbor's any good. How, then, does one go about "working the land"?

Clearing the Land. Let's start with a field that has lain fallow for some time. The first thing to do is walk it, checking for any enormous rocks or bushes or saplings. Rocks don't grow the way trees do, but they have an uncanny knack of making their way from the subsoil to the surface. In an old field you may find them there simply because the previous farmer knew exactly where they lay and decided they were either impossible or too much work to move. You may reach the same conclusion. But you have to know their location in order not to break your plow. Plows are strong. Still, they do break. And, believe it or not, some fair-sized rocks honestly weren't there the last time the field was plowed. They've surfaced through frost-heaving, leaching, or other soil conditions. These are usually fairly small and manageable. Harvest any over the size of a cantaloupe.

Bushes or saplings that have established themselves in the field must be cut down. Cut them as close to the ground as you can, and always cut them flat across. A sharp, spear-pointed sapling will go right through a tractor tire, and changing tractor tires is no great joy. It's also an unnecessary extra expense. Stumps from large saplings must be pulled out. Then with the field cleared you're ready to plow.

Plowing. The ground should not be too wet or too dry the day you plow. If it's too wet, the furrows will harden like rock pillars and the soil granulation will break down, reducing soil quality. If it's too dry, you probably won't even get the plow in the ground, and if you do manage to, the soil will break into large clods while the plow skips around. The time for plowing can be learned only by looking at and feeling the soil. The old hands at it know when it's just right, so plow when you see your neighbors revving up their tractors.

You start at the midpoint of one end of the field, not at one side. A tractor will buck a lot compared to a car, and your first task is to learn how to run it smoothly. Also, southpaws please note, almost all plows are right-handed. That is to say, the moldboard lifts the soil up and turns it over to the right. As you run the tractor down the center of a field, the trick is to get the plow to dig in consistently at the same level, usually about five to eight inches deep.

Standard plowing practice calls for turning the soil over completely. That's the way it's still done in most areas. However, the major criticism against moldboard plowing—and it's been a controversial subject ever since E. H. Faulkner's classic *Plowman's Folly* was published in 1943—is that by completely burying the surface trash, the organic material is blocked off from its oxygen supply and thus does not break down in a way most beneficial to the soil.

Disc plowing avoids the problem, chopping up the trash and only half burying it. But since the disc plow is simply not a practical investment for the small farm, what can you do? The answer is, plow so the furrow is but half turned. This will bring up the subsoil, at the same time only partly submerging the trash. With cornstalks and sod you will have to turn the soil over all the way, however, unless you have a disc harrow to chop them up before plowing. The trash, contrary to its name, is essential to your soil.

Turning your furrows half over may pose something of a social dilemma if your neighbors invert their furrows completely. The best way to lose your neighbors' help is to tell them how to plow their fields, at least your first year. Wait. Learn to plow their way. Then, some other year, figure out how to adjust the moldboard so it doesn't completely turn the furrow. Chances are for the first couple of years you won't be able to turn a furrow properly anyhow. This isn't to discourage you, it's simply a fact. You wouldn't try, say, flying without lessons, so it shouldn't come as a surprise that you'll have to learn how to plow— on the other hand, when you make a mistake on the field it might decrease your yield a little, but you won't end up splattered all over the countryside.

All right, there you are running down the center of your field, trying to keep the furrow straight and the plow set properly at the same time. Suddenly you're at the other end of the field. You stop. If you plow all the way to the end, how do you turn around? If you don't, what happens to all those irregular arches you've plowed into the field while trying to turn on it?

For the purpose of plowing, fields are customarily divided into two distinct sections, the lands and the headlands. The headlands are strips about fifteen feet wide at either end of the field which you leave unplowed as turn-around areas. The lands are what you plow first.

The reason "lands" is plural is that a very wide field is broken up into separate plow areas. A field up to a hundred and fifty yards wide you can treat as one land. If it's three hundred yards, divide it into two separate lands and plow each as an individual field. This is so you don't spend too much time going back and forth across the headlands.

So you've plowed one furrow and managed to turn around. Which side of the furrow do you go down? If you go down the right-hand side —remember, a plow is right-handed—you'll be scooping more soil away from the first furrow, leaving a deep trench down the field. Since you're not trying to build trenches but plow a field, how about going down the left-hand side? But then you'd be piling soil up against the first furrow and you'd end up with a ridge down the center of the field. You really don't want a ridge either.

Not all problems have a solution. In this case, you opt for the ridge. Every land has one down the center, and since it's the backbone of the field, at least visually, it has acquired the name "back furrow."

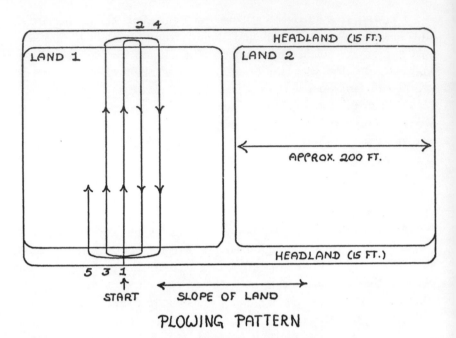

PLOWING PATTERN

After you've gone over the field twice and formed the back furrow, you always keep left of the previous furrow. In other words, you're plowing up the field, crossing the headland, and plowing down to the other headland in an ever-widening parallelogram.

When you've finished plowing one land, if your field is big you go on to the next land. Note that it is the width of the field that determines the number of lands, not the length. The length doesn't matter. In fact, the longer the better; it saves time. The purpose of lands is simply to avoid spending too much time crossing the headlands again and again.

If you have more than one land, you'll end up with the ditch between them that would have been down the center of the field instead of the back furrow had you made a right turn instead of a left one when you first started. It can't be helped. This ditch is known as the dead furrow. The big difference you've made by turning right while plowing is that you end up with a dead furrow only between the lands. If you plowed a whole field making left turns, you'd end up with nothing but dead furrows. Remember, the plow is always a righty.

Once the land or lands are plowed, you go across the headlands, making furrows at right angles to those of the lands. Then sit back and let the field dry for a few days.

Liming. If you're going to lime your field, after the field has dried a bit is the time to do it. Fill the grain drill with lime, set the adjustment for liming, and run up and down the field, starting one spreader's width away from one side and working your way over to within a spreader's width of the other. Then sweep once around the edges of the field.

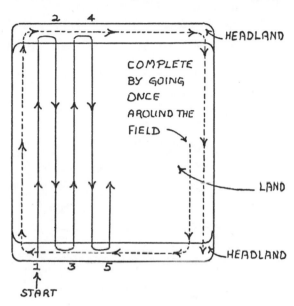

LIMING AND FERTILIZING PATTERN

Harrowing. So you've plowed the land. There are now lumpy furrows running up and down your field, maybe tousled a bit more from working lime into them. Looking at those furrows, you decide that a seed only an eighth of an inch or so in size will get totally lost. Right you are. To prepare the field for seeding, you will have to harrow it. That is to say, you want it to be level and smooth again, with all the lumps broken up.

If you're beginning to wonder why you should plow at all and then return the land to its preplowed state, well, that's not actually what you're doing. The soil will have been turned and broken up. Any organic matter will have been worked in, and what you end up with is a deep layer of loose, well-aerated earth. Don't plow, and you'll have an asphalt-hard field.

The harrowing pattern is more complex than that for plowing. You'll be covering the same ground over and over again, and your first time around can indeed be a harrowing experience. First you run the harrow over the field the same way you plowed. This levels it a bit. Then, starting at an oblique angle to the edge of the field, you use a harrowing pattern, rather like drawing a series of superimposed squares and oblongs in simultaneously increasing and decreasing sizes without lifting the pencil. See the diagram below. As if that isn't enough, it's about as dusty as the Sahara in a sandstorm. If you've always wondered why so many farmers chewed tobaccy, it keeps that old spittle washing grit from between the teeth.

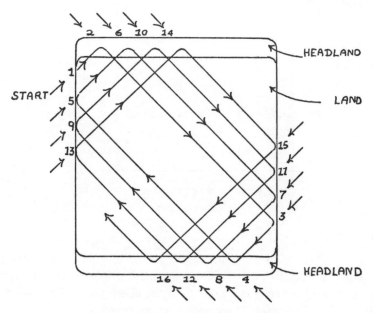

CROSS·HARROWING PATTERN

The harrowing pattern permits you to run the field continuously, and it cross-harrows the field so the clods of earth are attacked from different directions. Then you give it a final once-over at right angles to the direction in which the field was plowed and will be sown. This angle leaves the ground patterned so that when you sow you can see your tracks across it and know what ground you've covered and what you haven't.

Sowing. Unless you're broadcasting, load up the old grain drill with the seed you want and set the gears and holes for the appropriate number of bushels per acre. The sowing pattern is the same as that for liming. The trick is to run over the field once without overlapping seeded areas or leaving empty spaces. Don't worry, you'll do a passable, if not exactly professional, job the first time if you're careful and go slowly. If you err, it's better to err on the side of overlapping than leaving future bald spots. The second time you sow, your spread might be acceptable, the third, even good. The only way to get the hang of it is with practice, so don't fret over your mistakes.

TILLAGE MANAGEMENT

There are a few long-range factors to take into account when planning how to work your land. Knowing the lay of your land and mapping out your fields for their best use make good land management. Contour plowing, strip farming, and crop rotation are the three major principles to keep in mind.

Contour Plowing. Given a hillside to plow, the natural instinct is to go up and down. It's easier to work that way. It's also very destructive. Water at the top of a hill after a rain finds it easier to run straight down the hill too. And so it does. It joins the water that fell a little farther down the hill, and that a little farther still. Halfway down the hill it really begins to pick up momentum and volume. By the time it reaches the bottom of the field, you have parallel torrents running in rivers down each furrow. You lose not only moisture needed by the soil, but the soil itself, for the rushing water erodes the topsoil. Much valuable land was destroyed by runoff before the concept of contour plowing became popular.

In contour planting, furrows in hillsides are made with the contour of the land. It takes considerable planning to make each furrow level, but this eliminates erosion. Water seeps into the ground instead of sliding down it, and the topsoil stays where it belongs—on your field.

Strip Farming. An added boost to contour farming, where space permits, is to grow bands of different crops. The bands may vary in width, depending on the size of the field; for convenience, they're often the width of one land. This practice is known as strip farming because of the ribbon pattern that develops.

Strip farming is something you won't try right away, perhaps, but knowing about it makes good sense because the underlying principle is part and parcel of effective land management. While you plow and work strips for one crop, the intermediate bands, already sown for cover or bearing other crops, are left alone. These lands are then worked once the others have begun to take care of themselves. Sowing the intervening strips alternately to a cover crop that will be plowed under as green manure helps build up the organic quality of the soil.

Crop Rotation. Crop rotation brings to mind the by now almost trite maxim that you shouldn't plant land to the same crop over and over again because it will exhaust the soil. This is true, but it's not the whole truth about the benefits of crop rotation. Rotation plays a much more positive role than that. It improves farmland, not merely gives it a rest, and more importantly, it is one of the basic ways of controlling insects and weeds without dangerous chemicals.

Insects are for the most part useful creatures, and in many cases essential ones. Even those few considered simply pests in all probability serve some as yet unrecognized positive purpose. The whole insect problem has arisen chiefly because of crop concentration.

In nature a wild cornstalk might have been found growing every thirty or forty feet as part of a mixed meadow. Man has taken the plant and raised it in rows to the exclusion of everything else in that field. In the wild state an insect destructive to corn would find a plant, eat its fill, breed, and die. Most of the offspring would starve to death before maturity, simply because the next corn plant was too far away to find. Even if they found it, one more plant wasn't enough to feed them all. The same held true for diseases; with limited plant density, their spread was much less rapid.

But with a wealth of corn within munching distance, as it is under cultivation, living is easy for the bugs. Not only can all the offspring thrive, but their offspring in turn flourish. The old population explosion does its bit. Soon crops are eaten before they can mature, much less be harvested.

However, insects are neighborhood-oriented. Grow corn in one place the first year and on another field the next, and most of the insects won't relocate with the crop. A large part of your pest problem is solved by the simple procedure of crop rotation. The same holds true for weeds, some of which will thrive with one crop and be practically incompatible with another.

The other major benefit of crop rotation is that it builds up the soil. By alternating plants that use large quantities of nitrogen—corn, for instance—with those such as the legumes that actually add nitrogen to the soil, a productive balance is achieved. The same holds true for growing a green manure crop that is plowed under as part of the rotation scheme. It will build up the organic material in the soil, often at no loss of a crop, since your land should never lie bare. Planting a cover crop of, say, rye, for the winter not only protects the land from erosion, but also builds it up even further in spring when it is plowed under in preparation to sowing your cash crop.

A basic crop rotation scheme would have five equal-sized fields planted to four different crops at the same time. Five equals four because one crop—usually alfalfa—is grown on two fields at the same time. Let's say you want to grow wheat for your bread, oats for livestock feed and bedding, alfalfa for hay, and corn for the chickens. Now obviously, since you're using fields the same size, you'll have more wheat than you can use. But don't decide to cut down on the wheat crop. One of the basic rules of crop rotation is always to use the fields as laid out by the boundaries of the previous crop. If you have too much wheat, trade some, sell some, or use it as livestock feed.

TYPICAL FIVE-FIELD CROP ROTATION

Year	Field 1	Field 2	Field 3	Field 4	Field 5
1st	Oats * with alfalfa	Alfalfa	Alfalfa	Corn	Wheat
2nd	Alfalfa	Alfalfa	Corn	Wheat	Oats * with alfalfa
3rd	Alfalfa	Corn	Wheat	Oats * with alfalfa	Alfalfa
4th	Corn	Wheat	Oats * with alfalfa	Alfalfa	Alfalfa
5th	Wheat	Oats * with alfalfa	Alfalfa	Alfalfa	Corn

* As a nurse crop.

The following year you simply play musical fields. Each crop moves one field to the left, except the alfalfa, a good stand of which will last at least two years. The same goes for the third, fourth, and fifth years. By the sixth your corn will once more be growing where it was the first year.

The fundamentals of crop rotation can be listed quite simply, as follows:

1. The area for each crop should be about the same and the demarcation of the fields should remain constant over the years.

2. The whole cycle should contain at least one legume, one deep-rooted crop such as alfalfa, and one sod-building crop.

3. When possible, have one extra field in your rotation, or on the sidelines, so to speak. Keep this field in a pasture crop for two to four years, then reenter it into rotation with the grain crop. The field most recently used for the grain crop is then given over to the pasture crop for the same number of years.

4. By using one field for pasture, rather than cutting all the forage and hauling it to the barn, manure will be delivered directly to the field by the grazing livestock. This will assure maximum utilization of the manure, particularly the urine.

Strip farming and contour plowing may not be necessary on your field. Crop rotation always is. And although the lay of your land might not permit you to follow the above plan exactly, the closer you come to the ideal, the better your results. Sit down and plan your crops before you plow and plant.

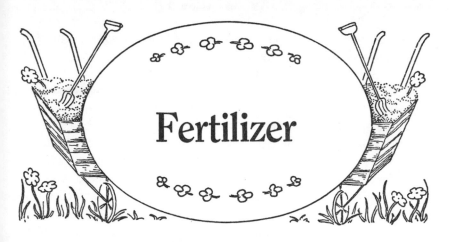

Fertilizer

Everything on a farm doesn't smell like roses. On the other hand, manure is nowhere near as odoriferous as the average city dweller thinks, and a well-made compost heap doesn't smell at all. Your farm needs both.

Almost from the time man first began domesticating farm animals, barnyard manure—the livestock's excretions mixed with bedding material—was used as a fertilizer to increase the yield of his crops. Some ancient agrarian civilizations, such as the Mayan, never went into animal husbandry on a large scale. It is now hypothesized that many once-great cultures were destroyed not by war, but by the exhaustion of the land surrounding their cities. The local agriculture could no longer support the urban population. All for the want of manure the fabled cities fell? Quite likely.

Agriculturalists used to believe that plants literally "ate" the soil. But with the birth of Justus von Liebig in 1803, the seed of radical change was planted in agriculture. Liebig's contributions to civilization were many and diverse, ranging from baking powder to chloroform. Among the discoveries backed by his chemical laboratory was the fact that plants merely extract certain substances from the soil, specifically, nitrogen, phosphorus, and potash.

From Liebig's new theory there were several corollaries to be drawn, and the nineteenth-century assault on traditional agriculture was soon in full sway. If you added enough of the right chemicals, you could even grow a bumper crop in sand—or, for that matter, in a water solution. The chemical industry rose to the challenge, and Malthus's dragon was slain. There would be food enough for everyone.

Unfortunately, as with all radical theories that presume to save the world by sweeping away the entirety of man's past endeavors, chemical fertilization and its adjunct, insecticides, did not turn out to be the panacea expected. Kept in its place, not all chemical agriculture is necessarily evil. But by overshadowing developments in the other areas of agricultural research, test-tube farming has done enough damage so that it could well take another century to reclaim once bountiful farm-lands from their artificially imposed fallowness. And the damage done by the insecticides in the long run will prove to far outweigh the good.

What happened was that chemical fertilizers alone were suddenly deemed the essential ingredients of agriculture. No heed was paid any longer to the tilth, or physical quality of the soil. What did the soil matter? Just add more chemicals if your corn wasn't growing and everything would be all right. The only problem was, as more and more chemicals were added over the years, the organic quality of the soil was lost. Once friable, rich earth turned into hardpan. The essential chemical ingredients were there all right; the high crop yields weren't. The reason? Simple. You can't grow abundance on macadam.

Finally, in the past couple of decades agriculturalists rediscovered soil—good, natural, organic earth, the way it used to be, and the natural cycle that nourished it. For instance, when you feed your goat hay, you're feeding it plants that took their nutrients from the soil. But the goat doesn't use all the nutrients in the plants. Save the manure and re-turn it to the field, and you're returning nutrients the soil can use to feed the plants to feed the goat. . . . Manuring, trash mulching, and plowing under cover crops are all ways of restoring the natural organic balance of the earth.

With that in mind, you'll no doubt use your manure rather than let it rot behind the barn. But you also want the best manure possible— yes, even the manure on your farm can be improved—and that takes a bit of planning.

BARNYARD MANURE

GETTING QUALITY MANURE

The kind of manure you get from your farm animals depends partly on the kind of livestock you have and what they've been eating. Animals on a good diet will make better manure. This is why it's so important to think of the farm as an organic whole, a small closed system operating on a cycle of its own. Every improvement you make at one point in the cycle will improve the following parts of the cycle. Feeding your stock the best will result in not only better meat, milk, or eggs, but better manure as well, which when spread on the fields will give better, more nourishing crops, which in turn . . . and on and on. But break into the cycle negatively—skimp on the feed, or let the manure go to waste—and you introduce a declining spiral of quality into your homestead.

In general, the quality of manure increases with the amount of protein in the animals' diet. More protein means higher-quality manure. This is the reason why horses, whose diet is strong on grain, produce one of the most valued manures.

Exceptions to this general rule are young animals, busy growing, and their mothers, busy making milk. They won't produce quite as nutritious manure as the rest of a herd even on the same good diet, because both lactation and body development use up larger quantities of calcium, phosphoric acid, and nitrogen. Technical considerations like these can't always be taken into account, obviously. If you're raising goats for milk, you want the milk; the manure is a by-product. You're not going to stop breeding the doe just to improve the manure. You'll simply plan on using more manure per acre during lactation than when the goat is in its dry period. If you buy manure, on the other hand, and are given a choice between that from dairy cows and that from beef cows being fattened, you'll want to choose the latter.

Buying or bartering for manure is not something one normally thinks of in the city. In fact, it's not something many chemical farmers would consider either. But it makes a lot of sense. If you look at your farm as the closed circuit it essentially is, you can see how bringing in manure from outside is adding quality. Whoever sells it to you is actually giving up a very valuable part of his land. Up to 80 percent of all the nutrients gathered by farm animals in foraging is expelled in their manure—the body can utilize only so much of the overall nourishment it receives. So letting manure go to waste is almost like strip-mining your farm.

Manure can lose nutrient value while still in the barn if precautions are not taken. As much as a fourth of livestock urine is lost because of floor leakage, considerably more, needless to say, in old barns with drains to remove the urine to cesspools "out back." A tight cement floor helps retain it. So does good bedding such as dry oat straw. Check the bedding twice a day. Gauge how much you spread out and how much it absorbed. What you want to strive for is balance: enough bedding to absorb all the urine and keep the animals clean, but not so much that the sheer amount of it gets out of hand, making the resulting manure mostly straw. Bedding contributes important nutrients of its own to the all-over manure. However, too much of it, or rough bedding such as cornstalks, may stay on top of a field long after it's spread, rather than breaking down into humus. When you're cutting your next hay crop later in the year, the old bedding will be mixed with the new hay, and you'll end up with a poor-quality crop. Also, of course, using more bedding than you really need is a waste.

STORING THE MANURE

Without proper care manure quickly loses many of its beneficial qualities. Nitrogen, phosphoric acid, and potash are lost through leaching, fermentation, and oxidation. These processes can't be entirely prevented, but with careful storage and protection the nutrient losses can be kept to a minimum. Ideally, spreading manure (except chicken) on the land and working it in immediately is the best preventative. The thing that militates against the small-scale farmer here is that in all probability he won't be keeping enough livestock to produce a quantity of manure sufficient to make immediate spreading worthwhile, except maybe in the vegetable garden. So he must build up his supply, from his own barn and/or his neighbor's. In either case, he'll want to know it's kept well.

The richer the manure, the more rapidly it will lose its value if not carefully stored. Proper storage means protection from the elements. The old way of piling droppings out back in scattered dumps is definitely out. If you must store the manure in the open, it should be collected in one heap, on a slightly concave concrete or hard-packed clay floor to minimize liquid loss. Pile it high, pack it tightly, and make the sides steep. Dent the top a little, forming a depression to collect rain water. This will prevent the rain from eroding down the sides and washing away nutrients. At the same time, it will keep the manure moist. In dry, windy weather you'll have to wet down the heap somewhat to prevent burning, or overdrying of the manure. With overdrying, it turns white, indicating there has been considerable nitrogen loss and even some shrinkage of organic quantity.

The best way to store manure is in covered concrete pits. Again, the manure should be packed tightly and moistened when necessary. Thus cosily stored, manure will keep well for months. The pit has the additional advantage that it can be used for the production of rotted manure if you happen to have a bumper crop of manure around sometime. Rotted manure is simply aged manure, usually that which has been stored over four to six months. The rotting reduces the total weight of the manure by almost half, but since the weight is lost much more rapidly than the nutrients, well-rotted manure is still good manure, merely in more concentrated form. This means less of a load to haul, and some farmers with plenty of manure around prefer to work with it rotted. Also, in the case of some plants, fresh manure will cause fertilizer burn. To avoid this, well-rotted, or aged, manure is used instead.

SPREADING THE MANURE

If you've stored manure well, it can be spread on the fields at your convenience. Often this means you'll be spreading it in the wintertime, when other work to be done on the farm is at its nadir of the yearly cycle. Spreading manure in winter is fine, so long as the fields aren't covered too deeply with snow, since that would leave the manure exposed to weathering.

A manure spreader can cover in two hours a field that would take you two months with a shovel and wheelbarrow. So it's a real necessity. You can pick up an old out-of-date model very reasonably at a local farm auction that will still serve the small farm well. Till such time as you can buy one of your own, borrow your neighbor's. If you're buying or bartering manure from him, usually you'll find he will also help you spread it. Or perhaps you can have your spreading done by what's known as a "custom man." The custom man has been a rural tradition ever since modern labor-saving devices came to the farm. In most small-farm areas you find farmers with more machinery than their neighbors who want to maximize the utilization of their equipment and thus will offer to use it on their neighbors' land for pay.

The amount of manure to apply to your fields varies with the crops you plant. Truck crops may need thirty tons or more per acre. With hay crops, one ton per acre will produce a good increase in yield, two tons will almost double the yield, three tons will do even better. But there can be too much of a good thing. If you apply eight or ten tons of manure per acre of hay, the nitrogen concentration may become high enough to cause lodging. That is to say, too much nitrogen weakens grass stalks, and the hay mats down to the ground, making it difficult, if not impossible, to harvest. Lodged hay, whether caused by excess nitrogen, a heavy rain, or hail, is not a total loss, since it can be turned over as green manure. But it was a hay crop you were after. Four to five tons of manure per acre per crop is a good average for most soil conditions. Find out what your neighbors are spreading—especially the one with a farm you admire—and what they think of your acreage. They won't be able to give you absolute figures to rely on, but between their ideas and those of your county agent or local agricultural experiment station you'll begin to have a feel for what's needed.

All right then, you may be thinking, if it's going to take *tons* of manure to fertilize the fields, it's hopelessly out of the question. After all, you were planning on a couple of pigs and goats, or maybe just a few chickens. They'd get you nowhere. Or would they? Four growing pigs will give you almost twelve tons of liquid and solid manure a year. Three goats will give you four tons, thirty chickens two tons. The combined amount of manure from even relatively little livestock is often enough for the acreage in actual use on a small farm.

GREEN MANURE

Not all manure comes from animals. Manure is anything that through its decay introduces organic matter and nutrients into the soil in compensation for that removed by crops, livestock, and the elements. Green manure, a crop raised for the express purpose of being plowed under, is as much a fertilizer as barnyard manure; it's just a different kind. Plowing under returns the organic matter of the crop to the soil. In the case of legumes, the most desirable green manure crops, the soil is given additional nitrogen as well. Most plants get their nitrogen from the soil, but legumes, with the aid of certain specific bacteria, have the ability to extract nitrogen from the air.

Green manure is a work-saver as well as a fertilizer. A large part of the organic material derived from green manure is in the form of decaying roots. Alfalfa roots, for instance, grow four or five feet deep into the soil. When the plants are turned over, the roots break down, or decompose, into organic matter at their tips as well as their tops. So water retention and soil quality improve all the way down. To plow those four or five feet deep yourself, you'd have to use dynamite. It's literally been done.

The amount of organic matter retained by the soil is obviously not equal to the total volume of green manure plowed under. For one thing, as the crop breaks down, it shrinks. In addition, during decay some of it disappears into the atmosphere as carbon dioxide. Then again, how much organic matter is retained depends in part on the initial quality of the soil. Loamy and clay soil will retain more than sand. Temperature is also an important factor. During hot weather more of the potential organic soil-builder is lost in the form of carbon dioxide than in cool weather. The worst possible combination is sandy soil and tropical heat.

Green manuring will not convert your acreage into a wonder field over one winter. Very little permanent organic matter is added to the soil by plowing under. Even with the best of conditions it may take almost half a century to double the organic matter in the soil. But done wherever possible every year, green manuring will prevent the deterioration of soil quality while improving it enough to increase the yield of the cash crop you plant next on the field by anywhere from 10 percent to 100 percent. A cash crop, incidentally, isn't necessarily one you sell for cash. It's simply a crop that's grown to be used rather than turned over into the field.

Organic matter in the soil is as much a *process* as it is a substance. That is, the decay of matter is as important as the material itself, for it is this process of decay that releases the nutrients. The different stages of decomposition give the soil different qualities. If a green manure crop plowed under remained as whole dead plants in the ground, the soil would resemble caked clay with hollows and layers of straw. Not conducive to good crops. It is the slow blending of decaying matter, the moisture it retains, and the organic material that give the soil the desired quality. The decay of the organic material continues till there is nothing left. If no new organic matter is introduced, only the claylike quality of the soil remains—or the dustlike quality, as it loses its ability to hold sufficient moisture. This is why organic matter must be continuously returned to the earth. And why it takes so long to build rich soil. For even as you turn over your green manure crop, you're planning what to sow as a cash crop on the same field the same year, a crop that, were man not farming, would also have remained to nourish the soil.

Both grasses and legumes can be used for green manure. Legumes, however, are preferable because they add nitrogen to the soil as well as organic matter.

Inoculation. Legumes must be inoculated before being planted. When I was a kid I thought someone sat around the old seed company with a hypodermic needle giving each little seed an injection. That isn't quite the case. Legume inoculation is done with the suitable culture of single-celled bacteria for the specific seed involved, but on bushels of seed at a time, not individually. The bacteria for clover won't work on alfalfa, or even sweet clover. And don't let the word "bacteria" scare you. If the clover variety won't touch the alfalfa, what do you think the chances of its liking you are? Nil.

Inoculation refers to exposing the seeds to the bacteria, using commercially available cultures, so each seed becomes coated. The seeds must be inoculated the same day they are to be planted, as the bacteria die when out of soil. A culture usually comes in one-bushel size—a plastic baggie with about a cupful of what looks like black soot. Mix with water according to the accompanying instructions, toss into a seed inoculator, or a home-made barrel version, with a batch of seeds, and crank like old-fashioned ice cream until the seeds and inoculating culture are well mixed.

A field itself can be inoculated through previous plant growth. For instance, soil upon which inoculated soybeans have once grown will retain the bacteria long enough to inoculate a subsequent crop of soybeans even if sown a year later. The same holds true for most legumes.

Inoculation is not a preventative against anything. It's a booster. When you want a solution to crystallize, introducing a few ready-made crystals speeds up the process. The same kind of thing is true of inoculation of legumes. You supply nitrogen-fixing bacteria. As the plant develops roots, the bacteria enter the root hairs, immediately forming nodules that become regular little nitrogen factories. The plant

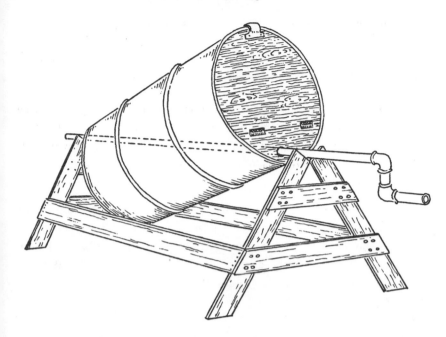

SEED INOCULATOR AND GRAIN TUMBLING MIXER MADE FROM OLD BARREL, USING CRANK AND AXLE MADE FROM PIPE AND HINGED WOODEN COVER CLOSED BY CATCH

supplies the bacteria with nutrients, the bacteria supply the plant with nitrogen. A very productive symbiotic relationship that might still have developed had the inoculation not taken place, but to a much lesser extent and, in the case of a field where that specific crop had never grown, perhaps not at all.

CROPS FOR GREEN MANURE

LEGUMES

Sweet Clover. First on your list, especially if you have bees. One of the hardiest of the clovers, this one comes predominantly in biennial form. Will do well in any part of the country where rainfall exceeds eighteen inches, well cycled throughout the year. It can be grown on most types of soil as long as there has been adequate liming. For the bees' honey harvest, let the crop stand till after the blossom stage before you plow.

Ladino Clover. This clover is a rapid-growing perennial two or three times the size of common clover. It does best in the temperate climate of the northeastern states, but is also grown in the Corn Belt. It does not make good hay; however, it is excellent for turning over.

Alfalfa. A deep-rooted perennial grown throughout the United States. It fares particularly well in the Midwest and Far West. Most soils will do, but extreme sandiness, clayeyness, or poor drainage will inhibit growth considerably. Good for green manure and hay. But the seeds tend to be expensive.

Trefoil. Comes in two varieties, bird's-foot and big trefoil, both native to Europe. Best adapted to a temperate climate, it is grown chiefly in the northern half of the United States, but can be cultivated anywhere. Big trefoil is the hardier of the two, so moisture-hardy, in fact, it can thrive under semiswamp conditions.

Nonlegumes

Rye. Your first choice of the nonlegumes. Because of its hardiness, rye is an excellent winter crop. That is to say, you can plant it in fall, at which time it will take root. Winter growth will be minimal but the crop will have a real head start come spring. You can plow under a winter rye crop as green manure before your regular spring planting. Soil requirements for rye are not stringent. Can be grown in any part of the country.

Redtop. The most widely adaptable cultivated grass in the United States. Grows well in all parts of the country except the extreme Southwest. Sometimes called "herd's grass" in the South. Timothy, on the other hand, is sometimes called "herd's grass" in the North. Same herd, different grass.

Timothy. The most important hay grass in the States, despite the fact that it is usually grown only north of the thirty-sixth latitude. A perennial grass growing in bunches, it will not grow in dry sandy soil, needing heavy earth and an abundance of moisture.

The planting and growing of these green manure crops are covered in the chapter on "Forage," except for rye and buckwheat, found under "Grain." For nonlegume green manure, stick with rye wherever possible.

PLOWING UNDER

Your green manure crop must be turned over at least two weeks —better yet, three weeks—before planting a cash crop on the same field. Otherwise its initial stages of decay will inhibit seed germination of the new crop and prevent plant growth. After two to three weeks, however, the decaying process will have progressed to the point where the soil is ready to nourish newly sown seeds. During warm weather, complete decomposition of the green manure usually occurs in about six weeks. Disc harrowing the plants two or three times before plowing them under helps speed things along. Cut up, the plants will decay more rapidly.

COMPOST

Your vegetable garden, like your fields, needs both animal and green manure. But plowed-under green manure doesn't really pay here, except as a winter crop to keep the land from lying bare. Plowing under is best left to your larger acreage of field crops. If you have a big vegetable garden—that is, you've gone commercial and are growing vegetables for market—you don't want to waste a cash crop by growing green manure instead. If you have just a small garden for your family, and maybe a roadside stand on the side, the dimensions of the plot won't be large enough to warrant the work of harrowing and plowing under. Then what? Your soil still needs organic material.

Composting is the ideal solution. You make your garden's own green manure in a compost pile, using all the excess vegetable matter on your farm—for instance, spoiled hay, plant tops from the garden, the straw from your wheat harvest, leaves from your orchard. When it's properly aged, spread it on the garden as you would animal manure. And lest you think that because you have a pile of decaying vegetable matter next to your garden the smell will be overpowering—it won't. A well-constructed compost pile doesn't smell at all.

Composting is a natural soil-building process that began with the first plants on earth and has gone on ever since. Leaves falling, ground cover wilting, trees felled, deceased animals and insects decaying, all that once lived gave up its life to the earth. Layers upon layers rotted— not dust unto dust, but life into new life, for from the decay new plants rose, feeding new insects and animals in turn, all in an endless cycle. The compost heap is nothing but an intensification of this natural process. In a good compost heap decay-causing bacteria have an almost ideal environment to multiply like gangbusters. Hyperactive, these bacteria break down the raw material of the compost so quickly that odors have no chance to develop.

INGREDIENTS OF THE COMPOST PILE

What's the best material to use for composting? Just about anything organic. The more variety, the better, because some trace elements (those of which only the minutest quantities are needed for plant growth) may be found in a weed more than anywhere else, or in a type of hay, or brewer's waste, or what have you. By mixing as many different things (organic, of course) into the compost pile as you have available, you can only help insure that the compost becomes more nourishing. Some of the more common ingredients are leaves fresh or old, hay, sawdust, wood ashes, garbage, nut shells, garden trimmings, manufacturing scraps such as dried blood, leather bits, sewage sludge, coffee and cocoa hulls, feathers. If it once lived, it can be reincarnated as a plant in your garden. But remember, some animal manure should be added to each pile for its nitrogen content. Chicken manure is excellent. If you have no livestock when you first start, tankage, bone meal, sewage sludge or dried blood available commercially can be substituted for the manure. If this gives you the idea that when you slaughter an animal you should add the blood to your compost pile—that's right. And when you're through eating, add the ground meat bones from the table. Use everything on the farm for something. Waste is just that: waste. The city cliff dweller has forgotten that garbage is an integral part of life, *not* something to throw away.

NO-SHRED METHOD

The speediest composting is achieved by shredding the materials used. They need not be ground up, but the smaller the particle size, the more rapid the process. A fallen hardwood tree in the forest may take a century to decay into humus. If that tree had been reduced to sawdust and mixed into a good compost heap, it would have yielded rich humus in less than a month. It's the same difference as that between putting a sugar lump and putting a spoonful of fine granulated sugar into your coffee.

If you can't get a shredder right away, however, you can still build an excellent compost heap that will yield humus in three to four months. A compost cage will help the process along and is easy to build.

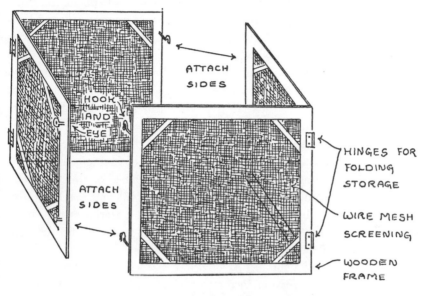

SECTIONAL COMPOST CAGE

Make the cage of wire screening so the compost can breathe. You'll be depending on aerobic action. (Anaerobic composting is not recommended for the small-scale operation since it takes more equipment for the best results.) For the same reason, the cage should be no more than five feet square. Any larger, and the center of the pile won't get enough air. You can make it as high as you want, but from the point of view of labor, five feet is about as high as you'll want to go. You'll also want to make the cage in two L sections so you can unhook the sides for easy moving.

A cage isn't absolutely necessary for composting. But besides the esthetics of a neat stack of compost as opposed to a heap, it has two very real advantages. For one thing, the edges of the pile are kept from blowing away. Sudden heavy winds can scatter a heap all over the yard. Of course, it will keep decaying wherever it lands, but that's not what you're aiming for. A more important contribution of the cage is that the sides of the pile are kept perfectly vertical. This prevents rain from leaching away many of the nutrients. Leaching can be minimized even more with shredding.

Floor your compost cage with half a foot of leaves, plant waste from your garden, spoiled hay, or garbage (organic, of course; no cans, for instance). Top this with a couple of inches of manure or used bedding from the barn, then about an inch of humus from your last compost heap if you have it, good bacteria-rich topsoil if you haven't. Unless you need particularly acid compost, for blueberries or the like, the final layer should be a thin sheet of ashes, granite dust, finely ground limestone rock or phosphate rock, to give the heap good mineral balance. Water the layer down, but don't compress it. Remember, the center of the pile has to breathe.

After you've watered down your first batch of ingredients, build a second layer onto it, again starting with half a foot of organic material and progressing from there. Keep adding layers till you reach the top of your cage. You should have plenty of organic material around even with only two or three acres of land. However, if you can't fill the cage all in one day, you can add to it as you go along.

The very top of your bin of compost material should be slightly concave so that a depression forms to catch rainwater. Even so, you'll have to check regularly to see that the pile stays moist. Should it begin to dry out, wet it down again.

A few days after completing your compost pile, you'll notice it shrinking, or settling. At its center the temperature is now reaching 150°F, so if you have any ideas about introducing earthworms to speed things up, forget it; they'd only end up being baked. Compost piles using only plant material, as opposed to ones to which manure has been added, can be used in conjunction with earthworms to make compost, but it's a slower process.

After two weeks of brewing, your compost is ready for a stir. Unhook the two sections of the cage and move it so that it stands empty next to the pile, which by now is holding its own shape. With a pitchfork peel off the top layer of the pile and toss it into the bottom of the cage. Then strip down the sides of the pile, about half a foot thick, and toss them in. The center of the pile decays the most quickly, so by making the outside the inside as you turn the pile, you get more even decay. Now, layer by layer, add the rest of the old pile to the new one in the cage.

Turn the pile again the same way after another three weeks, and a third time two weeks after that. In twelve weeks it's well-done.

SHREDDING METHOD

Shredding your compost material speeds up the process of decay to such a degree that this method is often called "two-week" or "fourteen-day" composting. And it really works that quickly.

Compost ingredients shredded before stacking decompose more rapidly for three reasons. First, although the original amount of material you're working with is the same, its total surface area is greatly increased. A piece of paper, for instance, has six sides. Tear it in half and the two pieces of paper together have twelve sides. Tear the two again and. . . . The multiplication of surface areas in shredding is phenomenal, and the surface area is where the decay-breeding bacteria live and work. This brings us to the second advantage of shredding. Small shreds let more air into a pile through the spaces between the solid material. More air means more bacteria can sustain themselves in their job of breaking down the compost. The third advantage is more moisture penetration, which also boosts the bacteria population.

Shredding compost saves the farmer work. You'll need a shredder, of course, but it's not the major investment some other farm equipment is.

By shredding the material before composting, with the subsequent increased bacterial action, you don't have to carefully build up layers of the various materials in your compost cage. Just shred the stuff, dump it in, and wet it down. Add a bit of humus on top of the pile to insure the introduction of bacteria—like a yeast cake to dough.

By the third day your compost heap should have begun to heat up. You can use a thermometer to test the inside of the pile, but why bother? Just stick your hand into it a ways. Thus you can check to be sure it's moist as well as warm. If it isn't warm, slide the cage off and turn the pile over into it, adding more nitrogenous material throughout. This is rarely necessary if you've added a good quantity to begin with.

Assuming the heap began heating on schedule, on the fourth day it should be turned. Moisten it down again after upending it. Be sure not to get it too wet, however. You don't want excess water to leach out nutrients and drain them off into the soil. It's not a bad idea to have your compost cage set on a sheet of plastic or a cement slab as an additional precaution against leaching.

The compost pile should be turned again on the seventh and on the tenth day. Moisten as necessary to keep it damp throughout. If all this turning sounds like a lot of work—it isn't. Shredded material is light and fluffy. You may not think you're tossing around pillows, but it won't break your back either.

By the fourteenth or fifteenth day your compost will be ready enough to apply on your garden. It will be rich, dark, and crumbly. Not as fine as pure humus, which is the next stage in organic decomposition, but fine to use. If you want potting-soil quality, let the pile age another one to three weeks.

But, you say, I intend to have a big vegetable garden. Certainly a pile of leaves and chicken droppings won't supply me with enough compost fertilizer. Well, let's see. Suppose you had a five-foot-square compost heap and piled it five feet high; that's a hundred and twenty-five cubic feet. The finished compost will have a volume of around ninety cubic feet. That's almost two tons' worth by weight. If you keep composting, spring, summer, and fall—say, for six months or so—by the end of this period you'll have made eight bins of compost. The total yield in good, high-quality humus will be over fifteen tons. Fifteen tons of humus a year will fertilize the soil in a very large vegetable garden and still give you some left over for the orchard and berry bushes.

As an interesting aside on what composting may hold for the future farm and self-sufficiency, anaerobic manure digesters have been designed that turn manure into humus. They operate in an airtight chamber, and as the manure breaks down, the liberated methane is piped off into a storage tank. From there it goes to the kitchen stove as cooking fuel. You can heat the house, generate electricity, cook your dinner, and still have organic fertilizer for your fields. There are actually some of these devices in use in Europe.

One English chicken farmer carries the process even further by having a chicken-manure digester in his car and a storage tank for the methane produced by the anaerobic action. He runs his pollution-free engine on the methane. Detroit would never buy the idea, however. A 1000-chickenpower Cadillac doesn't have the right sales appeal.

LIME

Like most sound agricultural practices, liming the soil was the custom thousands of years ago, particularly in ancient China. It did not become a general farming practice in the United States, however, until the twentieth century. This was partly due to the fact that the New

World's soil was virgin, and thus not in need of lime. Why bother when things grew as well here without lime as in the Old Country with it? But again, it was discovered that soil is not an endless reservoir, that it must be replenished. Today most farmers use lime on their fields, particularly in regions of heavy rainfall where it is easily leached away.

The amount of lime needed by the land varies considerably from region to region. Find a neighbor whose crops you admire and chat with him. Almost everyone loves to give advice. You don't have to take it word for word, but it will give you a good foundation for comparison.

The principal functions of lime in the soil are correcting soil acidity and supplying calcium as a plant nutrient. Soil that is too acidic does not readily give up its nutrients to plants; lime helps liberate those nutrients. If other soil conditions are good, the failure of such lime-loving crops as clover and alfalfa may indicate the need for liming your land. Of course, it's preferable to avoid the acidity rather than find out about it through crop failure. To this end, making a practice of liming all crops, particularly the legumes—excepting only the real acid-lovers like berries, radishes, watermelons, and potatoes—forestalls frustration.

The pH Factor. The degrees on a thermometer indicate temperature. On a warm spring day you can tell it's in the 70s; you're familiar enough with that form of measurement to guess pretty accurately within what range the temperature falls. There's no more mystery behind pH. It's simply a scale, like that for checking temperature, upon which to measure the acidity/alkalinity of a substance. In the same way that you can guess the temperature, a farmer with experience supposedly can taste and feel and look at a handful of soil to determine approximately what the pH is and what crops will grow well in it. As a beginner, you'll need a pH-testing kit.

The pH scale runs from the extreme acid of zero to the extreme alkaline of 14. Conveniently located in the middle is 7, which is neutral. The full range of zero to 14 includes much too extreme acid/alkaline conditions for most plants. Agricultural land usually ranges from a pH of 4 to one of 8, with most fruits and vegetables doing best in slightly acid 6.5 to neutral 7 soil, and legumes doing better in 7 to 7.5 soil.

Soil Testing. There's nothing wrong with the old tried and true method of tasting the soil to see if it's sweet and alkaline or sour and acidic—except that it doesn't give you an accurate pH reading, and unless you've eaten a lot of dirt and know what flavor you're after, it's going to taste, well, like earth.

You can still test the soil yourself, however, without having an educated tongue. Either with a specially prepared piece of tape, a variant of litmus paper, that when moistened and pressed into your soil changes color, or with a more accurate soil-testing kit containing reagents that will also tell you the nitrogen-phosphorus-potassium condition of your soil. A third alternative is to send a sample of your soil to a laboratory. Your county agent or local agricultural experiment station can either do it for you or refer you to someone who can. This takes time, however, and can be costly. You're probably well enough off with a soil-testing kit unless you seem to have real problem soil.

FORMS OF LIME

In order for limestone to dissolve, it must be crushed. Obviously, you can't put a one-ton block of it in the center of a field and expect it to dissolve like a sugar cube. But even thumb-sized pieces aren't small enough to use unless you're prepared to wait fifty or sixty years.

Lime for most soils is best applied in the form of ground limestone, at the rate of one ton per acre per year. Ground limestone does not break down immediately. It thus minimizes the dangers of overliming. The ground limestone acts very much like the twenty-four-hour cold capsule that releases its medication slowly over a period of time, only in the case of lime it's a twelve-month capsule.

Lime can be applied in various other forms, including quicklime and hydrated lime. However, quicklime, as its name implies, works quickly—often too quickly, burning your plants. Hydrated lime won't burn your plants, but it is so water-soluble that one good rain will wash it deeper into the soil than the plants can reach. Stick with ground limestone, or equivalents such as ground seashells or marl.

To give the farmer an idea of how quickly the limestone will activate, it is sold by screening size. That is to say, it is sifted through various-sized screen sieves, then graded. The screen size will indicate the size of the largest particles. Of course, many smaller sizes will also slip through, even pieces only 1/1000th of an inch in diameter or smaller. This limestone dust dissolves first, the larger pieces progressively later, thus maintaining the proper pH condition of the soil over a long period of time. Its overall period of effectiveness will be subject to such variables as types of crops grown and their effect on the soil.

The largest limestone particles passing through a 20-mesh screen won't begin to affect the soil for over a year and a half. Those passing through a 100-mesh—the larger the mesh number, the smaller the particles—will react almost immediately upon being spread on the ground. Since the finer the limestone must be ground, the more expensive it is, you'll want to use the largest possible particles that will be effective on your land. In most cases, 20-mesh ground limestone will suit your soil fine. Remember, the screen size indicates only the largest particles in the batch. There will be plenty of dustlike material in it as well.

SPREADING THE LIME

Limestone is best broadcast on newly plowed land and then worked in with a disc or a spring-tooth harrow. You will want to both plow and harrow anyhow, to prepare a good seedbed, so you only have to remember to lime in between the two steps. This general rule is not inviolable. Lime can be spread any time of the year that the field permits. Some farmers spread it before plowing, simply because that happens to be a convenient time for them. Others may spread it on top of winter grains after seeding, for the same reason. But the best is still the plow-lime-harrow-plant sequence.

How often you lime depends on the crop you plant. In the case of clover or alfalfa, both lime-lovers, some farmers lime lightly before each crop. On the whole, however, one liming a year, a ton of ground limestone per acre, should suffice. Follow the guidelines suggested by your soil test results.

The beauty of natural farming is that it's pretty hard to fertilize wrong. It's true that too much animal manure and lime can decrease your crop yield somewhat. The yield for one season might fall below expectations, but the long-range effects aren't destructive. On the other hand, chemical fertilizers and particularly chemical insecticides can literally lay the land barren if used excessively, and when they finally do wash away, not only is the organic quality of the soil they were used on destroyed, but also the chemicals continue their destruction all along the ecological path they follow, through your neighbor's land and watershed, the streams, the rivers, even in small part the ocean. Chemical farming helps destroy the world. With natural farming you'll help build a better one.

RAISING PH LEVEL OF A 5-8 INCH LAYER OF SOIL

Finely Ground Limestone (20-mesh or finer):
Tons Per Acre Approximately

	From pH 3.5 to pH 4.5	From pH 4.5 to pH 5.5	From pH 5.5 to pH 6.5
Warm-temperate area soil:			
Sand and loamy sand	0.3	0.3	0.4
Loam	0.5	0.8	1.0
Silt loam	0.9	1.2	1.4
Clay loam	1.3	1.5	2.0
Cool-temperate and			
temperate area soil:			
Sand and loamy sand	0.4	0.5	0.6
Loam	0.9	1.2	1.7
Silt loam	1.0	1.5	2.0
Clay loam	1.2	1.9	2.3

Fruit

All worms have an apple,
But not all apples have a worm.
—PETER REEVES

T HERE'S NO PLACE ON THE FARM quite like the orchard. For dew-fresh fruit, but also for picnics in blossom time, watching the bees gather nectar for your honey, and simply enjoying life. Making a detour on your way to feed the chickens in the morning and stopping by the orchard for a night-chilled Transparent apple is a country experience hard to pass up.

The only problem is, unless your land comes with an established orchard, it's going to take a couple of years' wait. That's why the orchard should be one of the first things you set out. If, as is likely, you move to the country in spring or summer, planning your grove of fruit trees right away and getting the trees in time for fall planting will give you a year's head start over putting it off till you're settled in. Even if you don't have time to get the trees, you can sow a green manure crop of rye to help prepare the land for early spring planting the next year. Fruit trees can be set in either season as long as they are naturally dormant when transplanted.

To speed up your first yield even further, try a few dwarf trees. You probably won't bother much with these once your big ones start bearing, but they will give you an initial crop to tide you over the second or third year of waiting.

ORCHARD LOCATION

The orchard is going to be around for some time. That's your first consideration in deciding where to locate it. You can put a vegetable garden on a slope that you expect to be part of a pond in a few years. To do so with the orchard would be a waste of either orchard or pond.

As a permanent addition to your homestead, the orchard should have not only good soil, but good air and water drainage as well. Thus a slope is the best location if you have one. Avoid low-lying sites, since these harbor the cold in winter, making any trees there more likely candidates for winter-killing than others in the area. A northern slope will delay blooming and subsequent fruit. A southern slope will speed up both. The ideal spot is on the small hills surrounding a valley or depression. Trees with early frost-sensitive blossoms can be planted halfway up the northern slope, the less sensitive trees halfway up the southern slope. No trees should be exposed to the windy hilltops and none to the frost-retentive bottom land. Rows of trees planted on hills should, of course, follow the contour system.

The soil in your orchard should be as good as you can make it. If you want to plant fruit trees right away and your soil is hard clay or sand, you'll have to build it up for each individual tree. It will pay you to fill the excavations in which the trees are to be set with improved soil, compost, rotted manure, ground rock phosphate, and rock potash to a depth of—hold onto the book—about five feet.

If you've ever dug a five-foot hole, you'll have some idea of what it's like to dig ten of them, or twenty, or however many as the fruit trees you're going to plant. The minimum would be two, since you should never plant just one tree of a given fruit if you want a good crop. There are two solutions to the labor problem. Either get someone with a back hoe or other mechanical excavator to come in and make holes for you. Or plant just two or three trees the first year, which will give you a good head start, and spend the rest of the year building up the soil for the others with cover crops of rye and clover alternated with liberal manuring. Another possibility is to set out a couple of dwarf fruit trees near the house, and leave the orchard be entirely until you've built up the soil for it thoroughly.

Of course, you may be lucky and have an orchard site with soil of good enough quality to use just as it is. What you want is a nice medium loam rich in organic material, but also with some sand. If you have straight clay or heavy sand, start digging.

PLANTING

Even if your soil is ideal, prepare to spend a bit of time with a pickax and shovel. The minimum size for a tree hole is three times the size of the root ball. In the case of fruit trees, the bigger the better. Usually one no smaller than three or four feet in diameter and two to three feet deep is dug. Pile the topsoil separately, since this is what should go back into the bottom of the hole along with well-aged nitrogen-rich compost and ground rock phosphate and rock potash. Don't use fresh manure. Spread most of the extra subsurface soil elsewhere and grow a cover crop over it.

If what you're removing is almost solid clay soil, and you're replacing it with the ideal light, humus-filled one, the improved area around the tree will act as a sponge. Water retention will be too much. In this case, put a tile drain at the bottom of the hole. This is simply a single row, or a cross, of sections of drainage pipe, usually four inches in diameter, spaced out on top of a layer of gravel across the floor of the hole. It leads the water away from the root area, culvert-fashion, to the surrounding subsoil.

The mechanics of planting fruit trees are important. All injured and broken roots must be pruned back. The roots are then spread out evenly on top of a layer of enriched soil replaced in the hole to raise the tree to its proper level. Make sure you keep the roots moist while you work. It's a good idea to mix up a bucket of mud slurry from your compost to pour over the roots when they are spread out. The slurry will coat the finer roots, keeping them moist, minimizing air pockets, and helping to settle them in. Fill the rest of the hole, tamp down the soil and give it a heavy dousing of water or light mud slurry. The tree should sit in the center of a slight depression about a foot in diameter. At the same distance out put up a two-foot-high wire mesh "collar" to keep out field mice, rabbits, etc. Outside of this mini-fence cover the ground with an inch of rotted manure extended all the way to eight feet

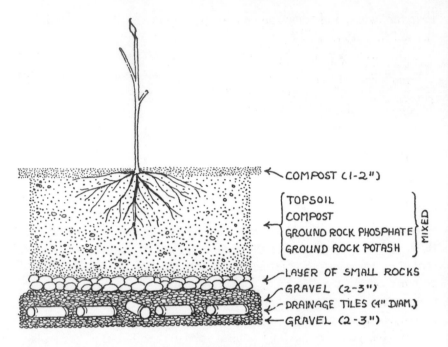

COMPOST (1-2")

TOPSOIL
COMPOST
GROUND ROCK PHOSPHATE
GROUND ROCK POTASH

MIXED

LAYER OF SMALL ROCKS
GRAVEL (2-3")
DRAINAGE TILES (4" DIAM.)
GRAVEL (2-3")

PLANTING A TREE

DIG A HOLE AT LEAST TWICE AS DEEP AS THE ROOT BALL OF THE TREE AND
ONE AND A HALF TIMES AS WIDE AS IT IS DEEP.   FILL AS SHOWN, MAKING
SURE TO SPREAD OUT THE ROOTS.

from the tree. Cover this circle in turn with about a foot of hay
mulch.

A regular fruit tree should be planted at least twenty feet from its
nearest neighbor. Dwarf fruit trees can be spaced as close as ten feet
apart. Set the tree straight to ever so slightly leaning in the direction of
any prevailing winds. The largest branch, like a weather vane, should
point into the prevailing wind. Prune back the branches of a newly
planted tree a little more, proportionately, than the root loss. This will
give it a chance to build a good strong root system. Give the trunk a
pole support to prevent wind whipping from loosening the roots.

One final thought on planting. Science is discovering more and
more about the roles bacteria and fungi play in crop growth. There are
no doubt some of both interacting with the orchard. An old farmer I
know would never plant a new fruit or nut tree without going around to
an orchard he particularly admired and, picking the best tree of the

kind he was planting, "borrowing" a shovelful of dirt as a "starter" for his own tree. Now I'm not saying this is necessary; the only reason I mention it is because his new trees never failed, and they were always covered in season with the most incredibly delicious fruit.

MAINTENANCE

Even organic gardeners spray their trees. Lest this make you sit up and wonder what happened to the natural approach, the spray referred to is not DDT or one of the countless other chemicals used by many commercial orchards. What the organic gardener uses is a miscible oil spray. It is a dormant spray, one applied before a tree begins its annual growth and which disappears before the leaves begin to emerge. Instead of poisoning insects, and the tree with them, a miscible oil spray simply puts your fruit tree in a temporary Baggie, suffocating everything that moves inside—except the tree, which is dormant. If it's not, well, you take it from there. Even miscible oil sprays call for caution and common sense in using, but, properly applied, they will not harm your trees.

Miscible oil sprays for dormant spraying are available in concentrated form through mail order and probably from nurseries in your area. Dilute and use according to the instructions. If you have an old established orchard on your spread and are trying to rehabilitate it, miscible oil sprays may not seem to be much protection at first because of the sheer number of bugs around. But have patience and eventually you will restore the natural balance. With new orchards and healthy, well-fertilized stock, miscible oil sprays should keep most of your problems in check from the start.

FERTILIZING

Early in spring, a month or so before a tree blooms, work into the soil around it some high-nitrogen fertilizer. A quarter of a pound of blood meal or eight pounds of manure for each year's growth is about right. Cover with a new layer of straw or hay mulch, keeping the area immediately around the base of the tree clear to discourage mice. For the little ring around the trunk, work in a two- or three-inch layer of well-aged compost instead. The compost should not actually raise the soil level at the trunk itself, or it may induce rot. A rule of thumb for young trees is to build the fertilizer layer out in a circle encompassing the widest branch of the tree.

TREE MAINTENANCE

1. CUT AWAY ROOT SUCKERS.

2. PRUNE CROSSED BRANCHES.

3. CUT AWAY WATER SPROUTS.

4. PRUNE UNDER BRANCHES.

5. FILL DEEP CUTS WITH PITCH TO PREVENT ROT FROM SETTING IN.

6. TO MINIMIZE INSECTS, USE A TANGLEFOOT BAND (A STICKY NO-MAN'S LAND THAT KEEPS DESTRUCTIVE BUGS FROM CLIMBING TRUNK).

7. FOR BEST GROWTH, MAKE A MULCH RING AROUND THE TREE, EXTENDING IT AT LEAST AS FAR OUT AS THE LONGEST BRANCH.

APPLES

There are over a thousand apple varieties, which gives you plenty to choose from. A few well-tested varieties, such as Gravenstein, Golden Delicious, Grimes, Rome Beauty, and Yellow Transparent, should be the basic stock of your orchard, but try some of the lesser-known ones as well. Apple trees tend to bring surprises, and a well-cared-for minor variety may give you the most wonderful fruit.

Apples will grow almost anywhere in the United States except in the hottest regions. They need the cool-to-cold winters during dormancy. Your local nursery will no doubt give you an indication of not only the feasibility of growing apples, but the best varieties for that area as well.

STOCK

Apple trees, except for the dwarfs, which bear small crops in their second or third year, usually won't give you any apples until their fifth to sixth year. But by the tenth year they're at peak production, yielding five to ten bushels per tree per year. And they will keep bearing for thirty years or more. So order two varieties at the very least. Not just because thirty or forty bushels of one kind might become a bit boring, but because some are eating apples and some for cooking. Also crossfertilization will increase your crop. Stock usually comes in one-, two-, or three-year-olds. You'll find the older ones more expensive, of course. On the other hand, being transplants, they are usually sturdier trees. Bought from a good nursery, two- to three-year-olds are your best bet.

PRUNING

If you buy two- or three-year-old transplants, they should need no pruning the first two years besides the initial one on planting to eliminate injured roots and take the wood down a bit proportionately.

After the first two years, your primary pruning job will be to make the tree easy to pick from and somewhat squat in shape. The center of the tree must be kept open; don't let the growth get too dense. Cut off branches that cross and rub against each other in the wind. Any branch so large it can't be pruned off with shears, should be cut in three stages with a saw, as in the diagram on page 64.

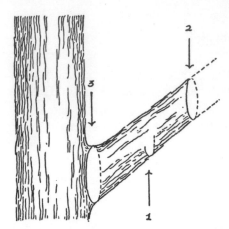

1. FIRST, CUT ONE-THIRD OF THE WAY UPWARDS THROUGH THE BRANCH ABOUT A FOOT FROM THE TRUNK.

2. NEXT, CUT THE BRANCH OFF ENTIRELY, WORKING DOWNWARD FROM THE TOP.

3. THEN CUT THE REMAINING STUMP OFF AS CLOSE TO THE TRUNK AS POSSIBLE, TO PREVENT ROT FROM SETTING IN.

HOW TO PREVENT STRIPPING BARK FROM TRUNK IN REMOVING LARGE BRANCH OF TREE

As the tree grows older, you will notice small, short cluster branches of buds developing along the real branches. These must not be cut off; they're the bearing spurs that will give you your fruit.

The spurs will bear lovely flowers—in fact, too many of them. If you walk through your orchard one day during blossom time and notice a host of pink and white petals falling, don't faint. It's quite natural. The tree would otherwise bear far too much fruit for it to carry and these dropping flowers are its natural way of cutting back the crop. Even so, once the fruit sets, you will probably have to tree-cull some of it. Don't let the young apples crowd each other. Not only can their weight break the branch, but too many apples spoil the crop.

PROBLEMS

Preventative care and healthy, well-nourished trees will minimize problems. However, there are some regions of the country in which cedar rust, scab, or fire blight readily attack apple trees. In these areas resistant varieties are grown. By taking the simple precaution of finding out from your county agent what the pests in your region are and selecting the appropriate resistant species, a lot of disappointment will be avoided.

The second problem-solving rule is, if it looks unhealthy, get rid of it. Branches with mildew or an area of prematurely yellow, withered leaves should be pruned out, and fallen spoiled apples or infected fruit removed. Don't leave sickly cuttings in the neighborhood of your trees. All diseased branches should be burned in a hot fire a good distance from your orchard.

The first week in June hang a trap on each of your trees for bugs. Use a wide-mouthed jar covered with a coarse wire mesh with holes just small enough not to admit bees. Fill it half full with a solution of one part molasses to ten parts water. A bit of yeast helps things along. Hang the jar on the southern side of the trunk. This will attract all sorts of nonsense. You'll even end up with a couple of persistent young honey-bees in the wrong pot. It will also, however, eliminate most of the apple flies, whose maggots have an annoying habit of tunneling into your apples. Keep the traps up for the rest of the bearing season.

On the other hand, if the answer to the old question "What's worse than biting into an apple and finding a worm?" * doesn't bother you, the bug trap may not be necessary at all. The easiest solution, of course, is to share the apples with your livestock—they get the culled wormy ones.

HARVESTING

Pick apples sun-ripe, naturally. Hold the apple in your hand, thumb against the stem, and with a light upward and sideways motion, twist it off. Be careful not to injure the spur on which the apple grew, since it will bear again next year. Also, don't twist the stem out of an apple you want to store; this would break open the skin to decay-causing bacteria.

You can't store summer apples, only fall and winter ones. However, there's no need to get a bellyache harvesting them. The cooking varieties make great applesauce and apple butter to raid the pantry shelves for later.

CHERRIES

Cherries come in two types, sweet and sour. Sour cherries aren't often seen in the city, but they can't be beat for preserves, jam, pies, and the like. They are also hardier than the sweet varieties.

---

* "Finding half a worm."

In general, cherries will do well under the same climatic conditions as apples. They are not as frost-fussy as apricots or peaches; in fact, they are one of the easiest stone fruits to grow, particularly the sour cherries. The best way to find out how they will do in your region is to check with the county agent and local orchards. Sour cherries begin to bear in their fourth or fifth year, sweet cherries two years later.

STOCK

Not only should you get both sweet and sour cherries, where possible, but you will probably have to get several varieties of the sweet. The really tasty ones like Bing and Napoleon are self-sterile and inter-sterile as well. That is, not only can't these trees fertilize their own flowers, even a neighboring tree of the same variety can't do it. They need Black Tartarians or other fertilizers around. To boot, the two "mates" have to bloom at the same time. Consult with a local nurseryman. This is nothing to try to coordinate through a mail-order house. If there's no nurseryman handy, make sure the trees you choose are double-bearing. These bear two different kinds of cherries at the same time. They have had a branch from a different species grafted to their trunk as a pollinating pal while still young stock.

A cherry tree, double-bearing or otherwise, will come grafted to rootstock better adjusted to supporting a yield than its original roots were. The two common rootstocks for grafting are mazzard and mahaleb. They are both wild cherry stocks, hardier than the superbred domestic ones. Mazzard stock is particularly good if you can get it. It is also more expensive, but in a long-range project like an orchard a little more money is well invested. Mahaleb-grafted cherries bear a year earlier, which often makes them the more popular. But the overall yield will probably be less.

Two-year-old grafted trees—the age is measured from the time of grafting—with the beginning of a well-spaced lateral branch system are the ones to buy. Fall planting is best for them. Mulch well for winter protection.

PRUNING

Like all fruit trees, cherry trees are pruned with the long-range intent of shaping them for easier picking and bigger yields. This means a tree open enough to permit light penetration. But sweet cherry trees naturally grow tall and upright. And you're working with nature; you can't change her too much and still expect cooperation. Therefore your sweet cherries will have to be pruned into taller trees than your sour ones.

Prune so that four or five branches off the main trunk become the primary bearing limbs of a cherry tree. They should leave the tree at a wide angle. The single top growing straight up is your leader. Leave it alone. But trim all other branches so they are shorter than the leader.

In deciding which branches to prune off, look for ones that point sharply upward. Also, you won't want the remaining branches to be too close together or all on one side. Mentally blow the tree up to climbing size. Could you climb it comfortably? A tree pruned for good climbing is a tree that will bear well. A good treehouse-tree, on the other hand, one that has three or four branches all coming out at the same spot up the trunk, is less apt to be a high-yielding tree.

Major pruning for shape need be done only once. Usually it's done when the tree is first planted, since this forces good root development. Maintenance pruning after that consists only of trimming out cross-branches that rub together, dead or sickly branches, and those that make a tree lopsided, filling out too much in one direction. Overall, up to ten major branches may be allowed to develop over the years.

PROBLEMS

Sour cherries are more disease-resistant than the sweet type—which is why you'll rarely find a sweet wild cherry. One pest to which they are both prey, however, is the tent caterpillar. If tent caterpillars strike, you have two alternatives. Let them be, and they'll eat every leaf in sight, multiply happily, kill the tree, and after several years decrease in numbers for lack of food. Or harvest the caterpillars as soon as you spot them. If you can manage to keep the population down, the cycle won't have a chance to start spiraling, and eventually they will disappear. If harvesting tent caterpillars is not your idea of how to spend a sunny Sunday, get some praying mantises to give you a hand.

You will occasionally find worms in your cherries. How you feel about the changing times and the "new squeamishness" will determine how you feel about the worms. My grandmother used to deal with worms in preserving cherries simply by going ahead and canning them. When you opened a jar of the incredibly tasty compote, you simply skimmed the worms off the top in the kitchen before ladling the plum-black cherries into the Limoges dishes for presentation in the dining room. The worms had been boiled out and were as neatly preserved as the cherries. Now I'd have to admit that finding worms in a can of commercial preserves these days would upset me a little, but not nearly as much as the DDT and additives in it that I don't see.

A final problem with cherry trees is birds. This one you're not going to avoid "noways, nohow." You'll begin to think you're running an aviary instead of an orchard. So you do one of two things. Either you generously plant enough extra trees to feed the birds too, including some mulberries, which they like best of all, or you cover the trees with bird netting. Myself, I prefer birds to ghostly draped trees that vaguely hint of nature gone away on vacation. Commercial orchards employ an automatic cannon that goes off with an explosive bang—but no ammunition—scaring the birds away. Again personally, I can think of better ways to spend my time in the country than being surrounded by cannon fire.

HARVESTING

The biggest harvesting problem with cherries is avoiding a belly-ache. If you work on the one-for-yourself, three-for-the-basket principle, you might survive. Pick fruit for eating without stems. They won't keep as long, but it's less of a strain on the tree. Fruit for storage or sale should be picked with the stems. Use a light, twisting, upward movement to separate the stem from the spur. Be careful not to injure or break the spur—it's the source of your future fruit.

PEACHES

Peaches present the paradox that they'll grow in almost any part of the country, but can be grown well in very few. Still, the smallest peach from your own tree will be tastier than most store-bought ones, even though peaches ship better than many fruits.

The peach needs both cold (below 40°F.) and warmth. Without a winter cold snap, the trees skip their dormant period and become too exhausted to bear. On the other hand, the early-flowering buds are very cold-sensitive. One frost and you're wiped out. And without summer warmth the fruit will not mature. Even so, the geographical range for homegrown peaches is almost nationwide, so try a few trees if at all possible.

STOCK

Since peach trees are so widespread but variable in adaptation, it's important to get stock suitable for your region. There are literally thousands of varieties, and although there is none best suited for downtown San Francisco or mid-Manhattan, there is a variety best for your farm. The trees are for the most part self-pollinating. But it's never a good idea to have less than two or three trees of any given fruit.

You'll get your first peaches after three or four years; the big yields will take another three. At that stage of the game you can count on four bushels per healthy tree. Spring planting is the best. A sandy or gravelly loam is preferred. Use a northern slope to delay blossoming if you're in an area of late frost.

PRUNING

Prepare to butcher your peach tree when you plant it. Peaches don't take too well to transplanting, so you will have to cut back the tops severely in order to encourage root development. Trim the leader back almost a third of the total plant height, making sure to cut just above a branch. The new leader will emerge from the junction and you don't want dead wood above it. Prune all the branches back to one and two-inch stubs. The effect you want is a spiked mini-flagpole. From the stubs eliminate all but three or four of the new buds that appear in the summertime. The object of the game is to develop a tree with three or four main branches rising together—in other words, a treehouse tree rather than a climbing one.

Regular spring maintenance pruning is the same as for other trees. Get rid of dead branches and shape the tree to be open and without crisscrossing branches. Also, when the tree begins to bear, you will have to thin out fruit growing too close together. One peach to every four or five inches of branch is plenty for it to bear. Even so, if a mature tree lets its branches droop heavily with fruit, you may have to support the branches with braces. Peach trees are prone to natural pruning, that is, branches break off from the weight of too much fruit.

PROBLEMS

Most peach tree problems will not strike a healthy tree. And those that are serious enough to destroy the tree won't be stopped by all the chemicals or anything else around, unless you kill the tree first anyhow. So sit back and let nature take its course. Fertilize your trees with nitrogen-rich compost in early spring to boost plant growth. Mulch the orchard, but keep the mulch at least two feet from the trunks to minimize peach tree borers. Prune away dead branches and those with injured bark, and pick off any strange bugs or their nests that you spot.

HARVESTING

Pick peaches when soft enough to give slightly under light thumb pressure. You've squeezed peaches at your local greengrocer's. Same principle, different fruit—the ones you pick ripe off the tree are much more nourishing. Twist fruit up and out as with others.

Peaches not fully ripe can be picked for storage if the season is running out on you, but be sure not to bruise them. Even gently harvested, they won't store for more than three or four weeks in a cool cellar. Make jam of the extras; peach jam rivals apricot for sheer lusciousness.

PEARS

Pears, like peaches, need to be winter-chilled. In general, they can be grown in the same regions as apples and peaches. They are hardier than peaches, but they flower earlier than apple trees, so watch for chills. Areas of consistent zero-degree winters are about the northernmost limit for pear trees, and with such temperatures winter-killing of the buds may occur.

Pear trees do well on poorer soil as long as it's well drained. In fact, very rich soil will produce overly lush growth, which in turn will encourage fire blight. Like Cassius, they should have a lean and hungry look.

STOCK

Self-sterile for the most part, pear trees are best ordered in different varieties if you want to be on the safe side, in which case they'll also need to be same-time bloomers. Buy one-year-old grafted whips around five feet tall. These should be about one-half- to three-quarters-inch thick, with smooth, even bark. Plant in early spring, on a northern slope where possible to protect them from premature flowering and a too-hot summer sun. Prune back about 25 percent on planting and don't use a fertilizer rich in nitrogen; this is one tree you don't want to grow very fast.

PRUNING

A minimum-pruning tree. After the initial planting trim, help shape it and remove root suckers when they form, but never give it a heavy pruning to boost growth. It would respond too well, and fire blight would follow only too often. Also very rapid growth makes the tree literally jump out of its skin; the bark splits, inviting borers and a multitude of other nuisances.

PROBLEMS

The main thing you have to worry about with pear trees is fire blight, and it may be hard to avoid in the healthiest of trees. Fire blight derives its name from the fact that the affected parts look as if someone had gone over them with a blowtorch while you weren't looking. It can strike all parts of the tree from flowers to main trunk, but not at the same time. This is your key to keeping pear production going. Trim off any blighted area as soon as you see it. Branches should be removed with a foot of healthy wood behind them. Don't handle the affected part and then the rest of the tree—would you blow your nose in someone else's old handkerchief? Keep all infected trimmings away from the tree trunk while you work, and burn them in a hot fire away from the orchard. The worst blight season is from bloom till fruit; this is the time to keep a sharp lookout for the culprit.

HARVESTING

Some pears must be harvested before they are ripe or they turn gravelly. The only way to find out if your variety falls into this category is to pick some when they are mature in size but just beginning to turn color, letting the others ripen on the tree. Compare quality. Next year you'll know. Pears picked just before the first blush of color can be cold-stored for two to three months or more.

PLUMS

Like most other temperate fruit, plums need to be winter-chilled in order to bear. They can be grown in just about any part of the United States where winter temperatures fall below 40°F. for a few weeks but not below −25°F. for the same period of time.

As usual, select a locally adapted variety, bearing in mind that prune plums will be needed if you want to make a lot of prunes. Regular plums can be dried too, but they have to be pitted and halved; their sugar content is not high enough to preserve them without fermentation.

STOCK

You will need at least two varieties, since most plums are self-sterile. All plums are root-grafted, which means on a one-year-old tree you'll be getting solid two- or three-year-old rootstock. Plant in fall if the winters in your area are mild enough so that temperatures do not go below the teens except on very rare occasions. Spring planting is fine as long as the tree is completely dormant, but fall planting is better where possible.

Space the trees fifteen to thirty feet apart, depending on the mature size of your particular species. Plum bark is sensitive to sun-scald (to a tree what a sunburn is to people), so wrap a couple of layers of burlap around the lower part of the trunk. Cut the wood back by 25 percent when you plant.

PRUNING

Pruning is mostly confined to shaping a wide, open, spreading tree, and, of course, regular maintenance trimming. Once fruiting begins, you may have to support some of the bearing branches like those of your peach trees. Eliminating overbearing by selectively removing small, unripe fruit will also be necessary. Keep all pruning cuts clean and, if possible, seal them over with pitch to protect against heart rot.

PROBLEMS

As with cherries, birds can be a menace. Try a mulberry hedge around your orchard. Birds simply love mulberries, and they will often stop at the hedge and fill themselves up before they ever reach your fruit.

HARVESTING

Harvest plums when fully ripe for eating, just a little before for canning. Prune varieties should be allowed to ripen till they drop—into a cheesecloth trampoline you have suspended beneath the tree. Remove fallen fruit daily and sun-dry on screen trays, turning occasionally. If it's rainy, oven-dry in single layers for six hours at 110°F.

Nuts

He that plants trees loves others besides himself.
—OLD ENGLISH PROVERB

A NUT TREE is one of the most valuable things your homestead can have, not only for the high-protein fine winter stores and excellent eating the nutbowl will provide, but for its highly prized timber as well—if it absolutely must be felled. The American chestnut is no more; the black walnut is not far behind in the race to extinction.

Walnuts are among the most valued of all temperate hardwoods, with a pair of good hundred-year-old trees worth more in dollars and cents than a whole furnished ready-to-move-into two-bedroom ranch house. The problem is it takes the tree that full hundred years to grow and only a couple of hours to fell it. There is some hope for the walnut, for people are beginning to wake up to the fact that even if they receive no direct benefit from a tree they plant, not even nuts for a dozen years or so, the world is that much better off for their having planted it. Granted there aren't too many people with this awareness yet—but at least there's hope.

The American chestnut situation is a little different. The reason the village smithy no longer stands beneath its spread, the automobile notwithstanding, is simply that the species has been almost totally wiped out by blight, as has the majestic American elm.

Had one fraction of the money spent on advertising by the industries most responsible for using up trees—the paper, housing, and furniture industries—been channeled into blight research and reforestation, these trees could probably both have been saved.

Be that as it may, not only the furniture and paper industries, but also you as a farmer owe nature a few trees. And your farm will only be the better for repaying a bit of the debt. Filberts will give you a much quicker yield than walnuts, but plant both where possible, along with as many other varieties as you can. It's pretty hard to harm anyone or anything by planting a tree, particularly a nut tree.

Because of the alphabetical order the following trees got themselves put into, the first couple may lead you to despair of getting any nuts at all. Please read to the end before you decide not to take even a nutcracker along to the country.

ALMOND

The almond, as you can tell by comparing one in its shell with a peach stone, is a close relative of the peach. Both belong to the greater rose family. Almonds come in bitter and sweet varieties. For the nut bowl the sweet are grown almost exclusively.

Although almonds will grow wherever peaches thrive, they bloom almost a month earlier and thus are often subject to spring chill. This means in many areas of late springs the tree can be grown only for decorative purposes. There's nothing wrong with that, however. If peaches grow, and if you have the space, why not try an almond tree?

Plant in early spring. Just treat it like a peach tree. Unless, of course, you want to heat your orchard the way some commercial growers do where the chill is too much for the blossoms. It's really not worth the work, however. Your almond tree won't miss the nuts much, even if you do. And on a northern slope the light will make it blossom a little later. Even if it's likely that your springs are too late, you might have an almond or two after all.

BEECHNUT

Here's a nut tree from which you'll never get many nuts. But it's a beautiful tree, and a boost to your wildlife. Plant a couple of extra beechnuts on your "back forty" for your children to make hand-hewn heirloom furniture from when they have a homestead of their own.

The tree is very hardy and will grow in almost any soil but that with poor drainage. However, since it is a taproot-dependent tree, you have to transplant it early. Also keep competing growth away from it for the first couple of years. Then you can let it go wild. Dormant oil spray in early spring, before the tree gets growing, is a good idea, but a hundred-foot-tall tree is a bit hard to spray, so you'll have to give up eventually.

If I had my way, everyone would be planting a beechnut, hickory, or walnut tree every year to celebrate their coming to the land. It doesn't take much time or money and would make all the difference in the world—to the world.

CHESTNUT

The American chestnut is dead, long live the Chinese chestnut. Although a tiny, unimpressive runt when compared with the majestic cathedral spread of the native chestnut, the Chinese variety yields excellent large nuts and can be grown in almost any part of the country. As with other early-blooming trees, it is often advisable to plant on a northern slope to delay flowering.

Several trees are needed to insure pollination. Of course, you want to plant them in the same area, not at different corners of your place. Twenty-five to thirty feet apart is fine. In some cases they will yield as soon as three years after planting, in most cases four years, and six years after planting you'll have a bumper crop.

Mulching and deep, fertile, sandy loam with plenty of organic matter are necessary. The Chinese chestnut wants a lot of moisture, but it needs good drainage too. The roots are more sensitive to standing water than those of most trees.

Pruning and care are about the same as for your fruit trees, except that you'll have to keep a sharp lookout for root suckers unless you want to grow a bush instead of a tree. Harvest nuts as they hit the ground. Don't let them mold, which they'll do quickly left to their own devices.

FILBERT AND HAZELNUT

Your northern slope—or eastern, it serves almost the same purpose —is getting crowded by now, because here's another one that should be planted there to delay blooming. Lest this make you think how bare your southern slope is going to look in comparison, that's the best place for your vegetable garden.

The difference between filberts and hazelnuts, in case you're pondering on the matter, is that hazelnuts are American while the filbert was imported from Europe. Filberts in most cases produce bigger and better nuts. Both are mostly shrubs rather than trees, although there are some fifty- to sixty-foot members in the clan.

Plant whatever variety is grown locally. Usually this means filberts in areas of milder winters and hazelnuts in the colder regions. Get at least two varieties for cross-pollination. Plant in early spring, using deep, fertile, well-drained soil. For bushes, filberts and hazelnuts have amazingly long and deep roots. They also have sensitive bark, easily sun-scalded and slow to heal. Bind the trunks up for winter in burlap to reduce snow-glare injury.

Root suckers must be removed every year unless you want a pincushion on your hands. For best yields, allow only three or four main trunks to develop. If you want to increase your filbert population, leaving propagation to the nut-burying squirrels won't suffice, although you'll see plenty of them around. Instead, let a sucker grow for a year on the parent plant. The next spring, take a knife and make a one-inch-long slit in the bark about six inches from the tip of the sucker. Then arch the sucker so the tip touches the ground. Stake it down so it can't pull itself loose. Bury the cut part of the sucker, from which roots will sprout. Use good compost, and leave the actual tip of the sucker exposed to grow as the plant. Water it well. Early in the second year cut the new plant free and transplant to a permanent location. You have just "layered" your first plant. Most willowy bushes and trees can be propagated in the same way.

HICKORY

There are several varieties here, and they're all cousins of the pecan. Treat the tree as you would a beechnut or a pecan. It's got a bear of a taproot, three to five feet long even on a young transplant, so be prepared to dig halfway to China when setting it in. Give it the best soil and compost possible, but no manure. It needs no pruning except to have the top cut back by 25 percent on planting.

If you hope to get good eating hickory nuts and not just a beautiful tree, get the shagbark variety. And if you can't get hickory stock—as is often the case because it makes a difficult and chancy transplant—find a supply of fresh-fallen nuts on your next outing in hickory country, and plant those. Plant plenty, for few will survive, *but plant!*

PECAN

Most of the above nut trees prefer the northern half of the country. Pecans are their southern counterpart and usually grow where cotton will. They need a long, hot growing season in order to yield.

Like all trees, pecans need deep soil, rich and well drained. And they have the deep taproot as well, which means you'll be digging again. Try five feet. Taproots on all trees must be buried absolutely vertical all the way. Don't try to rattail the tip into a curve or U-shape—you'd be better off saving yourself the time and trouble and throwing the tree away.

Cut back about 40 percent on planting. Fertilize well and protect the trunk from sun-scald. Very little pruning is necessary once the tree is growing, although usually the lower limbs are removed.

WALNUT

Grow black walnut trees for their beauty. Grow Carpathians for nuts. This hardy tree comes from the mountains of the same name and is thus very cold-resistant. It is also readily available through most nurseries—by popular demand. It's noted for its large yield of high-quality thin-shelled nuts.

Walnuts need deeep, well-drained, rich soil, of course, but a little acidity doesn't hurt. Plant as you do other trees. Protect the bark from sun-scald the first years. Black walnuts mature fully in a hundred and fifty years. But don't give up—the Carpathians will bear six to eight years after planting.

Berries

If ever I dies an' yo ain't certain I's dead,
Just butter some biscuit an' new made bread
An' spread 'em all over with raspberry jam,
Then step mighty softly to whar I am
An' wave dem vittles above my head,—
If my mouf don't open, I'm certainly dead.
—MONROE SPROWL

ONE OF THE TASTIEST SPOTS on your farm is the berry patch. Of course, it won't have just berries in it. Strawberries, raspberries, and blackberries, their names notwithstanding, aren't berries at all in the botanical sense. Grapes and currants, on the other hand, are. Just to balance things out, some with "berries" as part of their names really are berries—blueberries, cranberries, and gooseberries, for instance. Which somehow explains why they're all grouped together here, and certainly in no way detracts from their flavor.

BLUEBERRIES

To add further to the confusion of nomenclature, if you've ever had huckleberry pie, or picked wild huckleberries in the woods on a camping trip, chances are you didn't. Blueberries have small, soft, almost invisible seeds. Huckleberries have approximately ten big, stone-hard seeds. Eating real huckleberry pie is rather like chewing on a clam full of sand.

Blueberries are fussy about their growing conditions, requiring not only loose, acid soil, but a shallow water table and, at the same time, room for their sensitive roots to stay dry. They're very winter-hardy, however, making an excellent crop for the Alaskan homesteader. Not only do they like a cold winter—even the blossoms will withstand a 10°F. frost—but the cold is essential for most of the varieties to bear fruit. Thus as a rule they cannot be cultivated in regions where the temperature never goes down to freezing. However, minor species such as the Florida Evergreen, Rabbiteye, and Dryland, will grow without the cold, and sometimes in the more arid regions. They're worth an experiment if your land is down South.

PROPAGATION

Wild blueberries were probably spread from area to area by seeds in the droppings of birds and foxes. But don't sit around waiting for your local red-winged blackbird's help. A full-sized natural blueberry patch from seed is estimated variously to take between fifty and a hundred and fifty years. For best results on your farm, buy established blueberry plants grown from cuttings. Later you can multiply these by layering. But for the first few years your already started bushes will give you better berries, and sooner.

LOCATION

Full sun is what blueberries like best, as long as ground moisture is sufficient. If you have a swamp, blueberry bushes may be put to good land use there, but only on hummocks at least eight inches above the high-water mark. They love water, but their fine roots drown easily. The large quantities of acidic humus often found in swamps make excellent ground for blueberries because they provide the loose, open soil needed and often approximate the ideal pH of 4.5. If you have friable, well-drained soil in the garden, the plants can be successfully grown there as long as you work in lots of acid soil-builders such as pine and fir sawdust, oak leaves, or acidic peat moss several times during the year preceding planting. The pH must also be kept at 4.5 after planting.

PLANTING

Since you don't want to wait a decade for those blueberries, order three-year-old stock if you can. That's about the maximum age blueberries can be transplanted without a major shock, due to the extensive root system they develop. They give their first decent crop at around seven years. By buying three-year-olds, you'll have at least a small harvest after two years, and a full one of between ten and fifteen quarts not too long after that. Three-year-old stock will be sixteen- to twenty-four-inch bushes. Buying two separate species is suggested. Most blueberries are self-sterile, so cross-pollination is necessary for your crop.

You'll probably want to start the blueberry patch in early spring. Fall planting can be done in warmer climates, but in that case make sure you provide a very heavy winter mulch. Space the young stock with eight feet between them in rows eight feet apart. Dig up a three-foot circle for each plant and return to the hole a mixture of half soil and half acidic compost. Trim back the branches so they spread out no more than the roots do. Check that the roots have plenty of depth when planted.

MULCHING

Mulch when planting the first year with four inches of pine or fir sawdust that has been composted. If the ground is naturally acid, straw will do. The mulch should be renewed each year in spring, always extending the circle about one foot farther out than the widest branches.

PRUNING

Blueberries are quite prolific, and the new improved varieties are prone to overbearing. Too much fruit, particularly in the early years, can sap vital growing strength from the plant itself. For this reason some of the blossoms are usually trimmed off. If you have the patience to wait an extra year for berries, it's a good idea to trim all the blossoms off the fourth year, or, in other words, the first year for a three-year-old transplant. In subsequent years you'll probably want to trim off about a fourth of the blossoms. Just clip each blossom cluster, not the whole branch. One set of flowers left for every three or four inches is good spacing. They will become clusters of fruit, and you won't want to crowd them.

Suckers (straight shoots from the base of the plant) will develop, possibly even the first year after transplanting. For the first couple of years they should be pruned back to the base. Unless, of course, part of the main plant died in transplant, in which case trim away the dead wood and leave one vigorous sucker. When the plant is six years old, let up to four new suckers develop as part of the bush, but trim away some of the older branches that are bearing poorly. You don't want plant growth to stunt berry production at this stage. A good way to remember which branches are the unfruitful ones is to bend some colored plastic-covered wire around them while harvesting.

HARVESTING

The reason store-bought blueberries often have little more flavor than cotton is that they are picked as soon as they blue out, when they are still hard and easy to ship. Wait another week after blueing out to pick yours. This means you'll have to harvest them by hand, since a blueberry rake will gather the unripe with the ripe and may also bruise the soft, fully ripened ones. Take a berry on the bush between two fingers and twist lightly—if it falls off, it's ripe; if you have to tug for it, it's not ripe yet, no matter how blue it is.

BRAMBLE FRUIT

In the brier-patch category you'll find red raspberries, black raspberries, purple raspberries, yellow raspberries, and blackberries, including the dewberry, young, boysenberry, and loganberry. Take your pick. Personally, I would also include the wild rose, since the rose hips found on this bramble bush make excellent and highly nutritious eating—just remember to remove the hairy seeds, either with your fingers if you're eating fresh-picked ones, or by straining if you're making jelly or fruit soup.

THE WILD BRIER PATCH

In many areas of the country you'll find a bramble patch of native American blackberries or raspberries already growing on the farm. The problem then is to get a bit of organization into the thing. By hacking away large parts of the thicket you will be increasing the yield. What you want to do is clear paths through the patch so you can not only get at the berries, but prune and fertilize the bushes as well. You could, of course, try to organize the bramble into rows like a regular berry planting. For the average brier patch, however, it's better simply to make a cross through the middle with two paths five or six feet wide. This right away eliminates most of the center dead wood.

Remember to wear gloves and clothing that can stand up to the thorns when you tackle the job. Map out your projected cutting lanes with string or you'll have nothing but problems. Cut the plants on the path right down to the ground. Then dig up the roadbed and turn the soil. Those bramble-root stubs are going to be stubbornly anchored, so you'll have quite a job on your hands, but it will save a lot of digging later. Brambles are hardy as hell if you're trying to get rid of them. Cut to the ground and you'll have sprouts again next year. If in spite of your rooting job sprouts do come up in the paths later, be sure to dig them out promptly.

After you've cleared a central crossroads through the patch, you'll probably want to trim some of its outer edges as well to keep the rest of the canes in harvesting reach. The remaining canes, what's left of your patch, should then be cut back to four feet, with all dead canes being removed entirely, for more and better berries.

Now, having gotten the wild patch hopefully tamed and under control, give it a good dose of compost. A nitrogen-rich organic fertilizer will give it an extra boost with that first post-pruning crop. Mulching the paths fully will help retain moisture in the ground, previously shaded from the evaporative forces of direct sunlight.

PROPAGATION

If you don't have a wild brier patch around to convert into an instant berry garden, you'll probably have to get your initial stock from the nursery. You may want to do this anyway, since the varieties available today bear much larger fruit, including such novelties as the big, tasty amber and yellow raspberries. Buying from a nursery also permits you to get several varieties of one species, assuring a better crop through cross-pollination.

Buy the best two-year-old certified disease-free stock available and, unless you have plenty of space, stick with either red raspberries by themselves or with any combination of the other brambles. Red raspberries shouldn't be grown near any of the others, especially the black varieties, which are disease-prone and catch everything the red ones may be able to shake off lightly.

About 75 percent of your plants will take, and twenty-five bramble bushes will yield between fifty and a hundred quarts at maturity, which should give you an idea of about how many you will need. If you're having plants shipped by mail rather than purchasing them from a local nursery, their point of origin should be in your approximate latitude or have roughly equivalent temperature conditions. Were you to get your plants from a region much farther south than your homestead, their stage of growth would be too advanced and they would have a tendency, particularly the blackberries, to produce a lot of spindly canes rather than a more limited number of strong ones.

There are two types of bramble berry plants to choose from, the bush and the trailing varieties. Bush berries are hardier and easier to care for. Trailing varieties are more cold-sensitive on the one hand, more drought-resistant on the other. They can be trained on supporting trellises to make for easier picking—for both you and the birds.

The propagation of cane fruits is quite simple once you have a few roots and canes around. You can start a new plant with a root cutting from an old one. In the fall simply take a four-inch section of root about half an inch in diameter and bury it under two inches of sandy loam in a tray. Keep the tray in a cold frame if the winter temperatures do not go much below 30°F., in a cool, light spot indoors if they do. By spring the root piece will have sprouted. Presto, a new plant. Be sure to mulch it heavily its first winter outdoors. This is a fine way to extend your berry patch in a hurry.

Layering is another way. Let a few canes grow without pruning one year. By the end of the summer they'll be dipping down to the ground in a rattail fashion, with only minor leaf development along the tip. Anchor the tail with a split peg. Bramble plants take to layering so easily you don't even have to make the bark slit as with hazelnuts. Cover the part touching the ground with two inches of soil, making sure the leafy tip remains exposed. Come spring, cut the cane about ten inches from the anchor to sever from the parent plant. A new plant is all set to grow.

LOCATION

The best soil for bramble fruits is deep, sandy loam rich in organic matter. The plants will grow in sandy soil low in organic content, but the fruit will be small and seedy. Since the berries develop during what is usually the driest part of the year, the ability of the soil to retain moisture is a key to their quality. But any sandy soil can be built up with compost and mulching to induce a good harvest. Proper drainage is, as usual, central to success. Even claylike soils, with good dainage, on a slope, say, can foster excellent fruit if sufficient organic matter is incorporated for survival of the plant's hair roots. The bramble root system usually penetrates to the two- to three-foot level, so incorporate organic matter as deeply as possible before planting. A pH of 6 is preferred.

The raspberry in particular suffers from weak crowns. That is, the canes are attached to the crown (the point at which they emerge from the ground) in such a manner that they break off readily, especially when subjected to strong winds. For this reason raspberries should not be planted on an open, unprotected field, but rather at the edges where sunlight is profuse while wind effects are moderated.

Gently sloping ground is best. Choose a hillside that faces south, if possible, to give the bramble berries continuous summer sun all day long. If you're planning on several rows, contouring is essential to soil preservation. It might seem a good idea to plant alternate rows of red and black berries so you can conveniently pick both, but don't forget this would probably wipe out the black ones before they could even bear. Keep them apart—at least five hundred feet to be safe. Also, don't set out new plants in the vicinity of wild ones, again for reasons of disease prevention, nor, for the same reason, where potatoes, tomatoes, eggplant, or melons have grown in the past three years.

PLANTING

Bramble fruits are planted as early in spring as possible to give new growth time to strengthen sufficiently for wintering. The only exception is in the South, where the soil is warm enough for the roots to establish themselves in wintertime. The plants should be set out as soon as they arrive. Once leaves appear, the chances of success become much slimmer and the probability of disease increases immensely.

If your bramble plants are shipped to you rather than being toted home from a local nursery, the first thing to do upon their arrival is to make certain the roots are moist. Most shipments will arrive with the root systems packed in sphagnum moss and enclosed in plastic. Moisten if necessary and keep the plants in the shade until you get them into the soil.

Bramble rows should be spaced eight feet apart, with the plants three or four feet apart in the rows. Dig a hole for each plant one and a half feet wide by twice the depth of the root system. Fill the bottom half of the hole with well-aged compost. Each stem should be set with the soil level no higher than it was at the nursery. Spread the roots well before covering.

Settle the plants in with a heavy dousing of water. This will make them sink down so that they are below the general soil level. That's fine. The important thing is that the level around the plant itself be no higher or lower than it was before transplanting. If your soil drains poorly, it will pay you to ridge it so the whole row of plants stands slightly higher than the surrounding ground. Last but not least, trim all the canes back to eight inches.

MULCHING

The best way to maintain the moisture and humus content in your bramble patch is with a one-foot layer of straw mulch, which should be replenished yearly. Even so, if you are in a dry area, give the plants a good dousing of water just as the berries begin to form, for an extra boost. It's a good idea to mulch with six inches of used livestock bedding first and six inches of plain straw on top. Whatever you mulch with, don't bring it closer to the canes than six inches—to keep field mice in their place.

Nitrogen is also important to the bushes in fruiting. The addition of tankage or dried blood to the soil before mulching will supply large quantities of nitrogen. Two good handfuls spread around the base of each bearing bush once a year is about right.

PRUNING

The first thing to remember about raspberries and blackberries is that their canes are biennial. That is to say, the first year's growth does just that, it grows, producing no fruit. The second year these canes fruit. Meanwhile new canes are growing for the following year's crop. The old canes must be pruned off after harvesting, since they will not bear again. Without pruning, the yield of large berries will be decreased.

Pruning for the raspberries and the blackberries is about the same. For both, "heading back" is done very early in spring before any new growth begins. The number of young canes, those that grew the previous year to bear the current year, should be reduced to eight, or at most ten, the larger number for red raspberries. Cut off weak canes, and try for somewhat even distribution. Lopsided bushes make fewer berries. The remaining canes should be trimmed back to between four and six feet, according to how your particular bush bears. If it tends to carry its fruit close to the tips, the pruning must be moderate; if it carries fruit midway on the canes, the pruning is more severe, and so on. The purpose of this pruning is to reduce the number of berries borne, in turn increasing the size of the ones produced. You'll also want to prune the laterals sprouting from the main canes back to eight inches.

A post-harvest pruning removes all the branches that have borne fruit and, having borne once, will bear no more. The canes, being biennial, usually will die back anyhow. Your job is to cut them beforehand, so the plant doesn't expend useless energy on them. This also helps prevent the spreading of any disease. Trim the canes off at ground level; leaving a few inches of stubble invites rot and other problems.

In addition to the regular spring and post-harvest pruning, black raspberries must be kept in check through early summer. The reds you can leave alone. As a black raspberry bush develops new canes for the following year's fruit, these canes should have their tip buds cut whenever they reach the two- to two-and-a-half-foot length, to keep them short. This cutting, again, is to coax bigger berries from the plant. It will also make your picking easier the next year.

Another harvesting aid is tying each plant to a pole or, if you have a whole row of them, stretching parallel wires from poles at both ends of the row so the plants are enclosed between them. This will give you a neat hedge that doesn't constantly ensnare you in briery tentacles.

Trailing vines, of course, may be trained up trellises as they grow. The question on the small farm is whether this is worth all the extra time and effort. Pruning you have to do, but staking and training you may well want to let go till future years when you aren't quite so busy getting the place started.

PROBLEMS

Well-mulched, pruned raspberries and blackberries should give you no trouble if grown from healthy rootstock. Just remember to keep the raspberry reds and the rest away from each other. In part this is a preventative measure against anthracnose, a fungus disease characterized by rough gray and black blotches on the canes. Trim off affected canes as soon as you see those telltale blotches. As with all diseased prunings, they should be burned in a hot fire. The same holds true for spur blight, manifested by reddish-orange to purplish-gray spots at bud junctions, or spurs. More drastic measures must be taken with raspberry mosaic, in which the leaves curl and turn a mottled yellow-red. Dig out the affected plant and destroy it. Raspberry mosaic usually strikes when the rest of your bushes are a healthy green, so you shouldn't have much trouble diagnosing it. In fall, of course, all the plant leaves are bound to curl and turn color. Don't dig these plants out in a fever of doctoritis!

HARVESTING

Blackberries ripen black, but they're not ripe just because they've turned black. A ripe berry will fall into your hand after the lightest tug with two fingers, one on each side of the stem. Usually this is three to four days after they've turned black.

Raspberries are soft when they're ripe. You'll even learn to sniff out when a bush is ready for the berry pail. Harvest daily; they ripen quickly. If it rains during berry-picking season, make sure to get all the ripe ones right afterwards, or they will mold. Any moldy berries you find should be culled out; mold ripens even more quickly than your fruit.

CURRANTS AND GOOSEBERRIES

The *Ribes* family are hardier than almost all other fruit-bearing plants, which explains the popularity of these easy-to-grow bushes in northern Europe, where fresh fruit is often hard to come by. In America, on the other hand, the *Ribes* are not very popular. Which is unfortunate, because sun-ripened currants and gooseberries are fabulous.

Part of their sad lack of popularity in this country stems from the fact that the berries don't lend themselves too well to machine picking and the ripe fruit is hard to ship. Another reason is that the family is an alternate host for white pine blister rust and should not be planted near these trees; thus there are strict limitations on interstate shipping of *Ribes* in some areas.

But the first reason is a strictly commercial one, and the second can be easily taken into consideration unless your spread is very small. So why not some good *Ribes*?

PROPAGATION

Both currants and gooseberries may be propagated from cuttings. Currants take root a little more readily than gooseberries. For your initial order, particularly of gooseberries, the plants should be American varieties. European ones, the subject of more intensive breeding and care, give bigger yields, but there is often some difficulty getting them to take well here. Still, if your nursery stocks them, a few imported bushes are definitely worth trying too.

Gooseberries come in red, yellow, and green varieties. The green are actually a translucent white when ripe. The red are the sweetest. Personally, I prefer the green; they're the most tart, and make refreshing summer eating. Currants come in various shades of red, black, and white. Grow some of each. Incidentally, currants make the world's best homemade wine.

For gooseberries, order well-rooted one-year-old stock. Two-year-old stock is even better, but if you buy these you have to make sure you're dealing with a good nursery—which actually you should do in any case. Quick-buck operators will sell all their good stock as first-year plants, culling the others to let them grow to second-year plants. You get two-year-old stock, all right, but the worst of the lot. The same holds true for currants, although these, being sturdier, are likely to fare better as two-year-olds than the gooseberries.

LOCATION

Gooseberries and currants both prefer a heavy, even clay soil to a loose, sandy one. This is particularly true in more southern areas, where the heat retention properties of sandy soil can seriously impair the plant growth. Their roots are spreading, but quite near the surface, which intensifies the heat problem. On the other hand, this shallow root structure also permits the bushes to be grown in areas of too much ground moisture for other fruit plants, as long as drainage for the top six inches of soil is good.

Both plants have the added advantage of doing well on locations too windy for other berries, such as hilltops. In fact, areas exposed to a fair amount of wind are better for them, because good wind prevents mildew, an otherwise common problem, with gooseberries in particular. The bushes are so wind-hardy that they are often planted as a windbreak for other plants on an exposed side of a garden. Northern light is preferred.

Currants and gooseberries may also be planted in areas of partial shade where other fruit plants will not bear. This is an especially good idea in areas of hot summers where the tender surface roots might otherwise be scorched. Remember the mildew problem, however; the gooseberries will need good air drainage along with their shade.

One absolutely crucial factor is the location of any stands of pine, yours or your neighbors'. *Never* plant these bushes within a minimum of a thousand feet from any type of pine tree. White pine blister rust cannot spread from pine to pine, but put currants or gooseberry bushes together with white pines, and you risk not only wiping out the entire white pine population, but even affecting other varieties of pine. However, the spores cannot be wind-carried more than a thousand feet, hence the spacing rule, which, if followed, will adequately protect the local pines.

The soil for currants and gooseberries should have a pH in the vicinity of 7. Plenty of compost boosts yields. Lots of potash should be used, and some nitrogen, but not too much or you'll end up with more leaves than fruit. Therefore don't use chicken or rabbit manure; other types are fine.

PLANTING

Currants and gooseberries are very early spring bloomers. It's pretty hard to beat them to the gun. No matter how early in spring you try to plant them, you'll disturb the incipient flowers, so plant in fall instead. They lose their leaves and enter the dormant stage in early autumn, and thus the root system is given plenty of time to develop before winter's severity overtakes the land. Bushes five feet apart make a good planting layout, except for black currants, which are larger than the others. They're usually spaced seven feet apart instead.

Plant in soil that has been well mixed with potash-rich compost, including wood ash and bone meal, setting the stock slightly lower than they were in the nursery. If the roots seem dry, moisten them down before planting, although they're better off, of course, if not allowed to dry out in the first place. Trim off any broken roots. Pack the soil down fairly tightly around each plant. Trim all branches longer than ten inches.

MULCHING

Give the transplant plenty of surrounding mulch except for the field-mice circle of six inches' radius or so that you leave free where the plant breaks ground. The mulch must cover an area sufficient to protect the spread-out, shallow root system. A four-foot doughnut of it around each bush is good. Enlarge the doughnut as the plant grows.

PRUNING

The best time for pruning is early winter. First trim off weak branches and those lying close to the ground. In Europe gooseberries are sometimes shaped to form small trees eventually. This can often be done to good advantage, although the method is rarely used in America. Try it on one of your sturdier bushes and match the yield against the bush-pruned ones. In any case, gooseberry bushes should always be trimmed so there is an open space toward the top, where they sometimes tend to develop a tight head. The breathing space aids in mildew prevention.

Currants and gooseberries bear most profusely on their second- and third-year growth. Once your bushes are established, make a habit of trimming off all the four-year-old wood among the branches—you'll recognize it by its thickness. Remember to trim old branches all the way back to the ground. On a mature bush you want to leave four branches from each previous year's growth, except for a black currant, whose two-year-old wood should predominate. All these branches will have laterals of varying lengths. The laterals should not be cut back unless they've been injured. Trim for symmetry and you'll be trimming for best yield. All snippings, particularly those that don't look healthy, should be burned immediately. Never let them lie in the mulch.

PROBLEMS

White pine blister rust should be no problem if you plant your bushes the appropriate distance—a thousand feet or more—from any white pines in the area. The same type of anthracnose that affects the other cane fruits may affect currants and gooseberries; it is treated the same way as for the bramble bushes. Cane borers can be a nuisance. A cane that wilts early toward its tip is probably infested with them. Cut back until you find the culprits, and burn the whole cutting.

HARVESTING

The harvesting season for currants extends over a month or more. If you want them for jelly, do your berry-picking before they're quite ripe because the pectin content is higher then. Ripe ones for eating are, of course, picked much sweeter. Harvest currants in clusters, like grapes. Don't try to strip off the individual berries by the handful or you'll end up with pulp in your hands.

Goosesberries ripen almost a bushful at a time, which is convenient. You'll have to harvest only a couple of times in season, and once you've tackled this thorny job you'll be glad the ripening span isn't as long as that for currants. Some pickers don heavy canvas gloves and strip a whole branch at a time. This works fine on the greener berries wanted for gooseberry jam, provided you don't mind some pulpy squashed berries.

GRAPES

Grapes will grow in almost all temperate and many subtropical regions. European grapes can really be cultivated only in California and Arizona; the native American varieties, on the other hand, will probably thrive wherever you do. And well worth cultivating they are. Although I've never found grapes to rival some morning-chilled ones I picked south of Herat, Afghanistan—which certainly must vie with the ambrosia of the gods—almost any grape is a delight to the palate.

The three major species of grapes grown in America are the *Vitis labrusca*, an eastern variety of slip-skinned grape typified by the Concord, which is also grown in the west central region; the *V. rotundifolia*, a southern grape whose name rhymes with magnolia, typified by the muscadines, scuppernong, for instance; and *V. vinifera*, the European wine grape grown almost exclusively on the West Coast, typified by the emperor and Johannisberger Riesling.

For a permanent contribution to the world around you, grapevines are a good start. A well-tended arbor will outlive you by many years, giving pleasure to all. And for a personal experience that's hard to beat, take a couple of real friends, a wheel of homemade goat cheese, your own fresh-baked whole-wheat bread, and a bottle of wine from last year's crop out to the sun-drenched, fragrant arbor in grape-picking season and harvest the morning away.

PROPAGATION

Grapevines are usually propagated from cuttings or by layering. Southern muscadines are a little more difficult than other species. Whichever you choose, for your initial planting you should purchase one-year-old stock, from a local nursery, if possible, to assure easiest acclimatization. Surprisingly enough, a year-old vine with a good root system will fruit as quickly as a two-year-old transplant. What you are looking for is not length of the roots, but their number. You want a good, solid, bushy root system.

LOCATION

Grapes prefer very well-drained, warm soil rich in organic matter, and plenty of sun and air circulation. For these reasons they are often grown on hillsides, particularly those with southern exposure. If you have a sunny hillside, fine. If not, you can put your grapes on the south side of the house. But don't train grapes up the walls like ivy; this would increase the likelihood of early spring frost damage.

Open arbors are ideal. Face the open ends so that the prevailing winds run through the arbor, not up against it. In some areas arbors are used as summer shelters for tractors and other farm equipment. They suit the purpose, but I'd rather have the outdoors dining room. Although grapevines grow rapidly, you won't need to build an arbor for them until your second or third year. You will, however, have to stake out the plants with the future arbor in mind.

PLANTING

The best pH for grapes is 6 to 7, or almost neutral soil. Plant in the spring, but plow or dig up the land deeply the previous fall if possible. A cover crop of rye turned under before setting out the vines is an excellent way to boost the organic matter in the soil. Follow the rye crop with ground rock phosphate and granite rock or green sand spread liberally over the whole area to be used. If your soil is too acid, limestone will be needed as well.

For an arbor, plant the grapevines six to eight feet apart. For vines to be trellis-trained, ten or twelve feet is often better spacing if you have the room. You may think your young vineyard looks a bit sparsely settled at first, but two prolific vines have been known to cover a twenty-four-foot arbor all by themselves.

Pruning back grapevine roots to seven inches encourages the development of feeder roots. If you look carefully, you'll notice the vines seem to have a double set of roots: those just about at the soil line and a second pseudo-crown that spreads out below. The top ones are feeder roots; they run quite close to the surface, like those of your cane fruits. The second group of roots go quite deep, some to eight or ten feet. By forcing the development of the feeder roots and using rich compost and mulch, your plants are given a much better start. If there seems to be just one layer of roots to a vine, don't worry; in some cases it's very difficult to distinguish between the two types of roots. Plant the vine anyhow.

But first, while you're pruning the roots, cut back all the canes except two right down to the crown. Even these two should be pruned enough to leave only two or three buds on one and a single bud on the other, unless you want your vine to grow high quickly, reaching for, say, the top of an arbor. In that case, cut the two canes back to leave eight buds. As soon as the new shoots are half an inch long, prune off all but the top two.

Spread the roots out on a layer of topsoil in a hole a foot and a half in diameter and as deep as the longest root, making sure the bottom set is freely spaced. Tamp it down with some soil cover, then spread the surface roots, just as evenly. Cover with additional soil until you reach the level of the lowest cane bud. The new plant should sit a little lower in the ground than it did at the nursery. Help it settle in with a bucket of water poured slowly into the hole, making a damp ring around the transplant.

MULCHING

Mulch each vine with six inches of good compost, leaving the plant to grow out of a volcano-like depression. Don't bring the compost right up against the cane, however. For mature plants, half a foot of alfalfa hay mulch is excellent.

PRUNING AND TRAINING

Much is made of the mysteries of pruning and training grapevines. There are head-pruned, cordon-pruned, and cane-pruned vines, not to mention all the combinations thereof. Volumes have been written about them. Just to mention a few, there are the Chautauqua, Fan, Umbrella, Umbrella-Kniffin, Hudson River Umbrella, Single-Trunk, Two-Trunk, Six-Cane Kniffin, Munson. . . . Whichever do you use?

Well, why not simply forget about the whole complex matter? Unless you intend to establish a major vineyard, your grapes can do without all the fine distinctions.

Stick with the basic principles. First, do your pruning in late winter. Secondly, a vine will not bear a full crop till it's five years old, so leave it a little more permanent wood each year, building a sturdy vine. Thirdly, fruit is borne on the current season's growth, not the preceding season's, as with your bramble fruits. Therefore you're not going to reduce the yield by pruning back the canes each year. However, the buds on the previous season's canes are what produce the new bearing growth, so you must leave some buds. As a general rule, leave one bud on canes the diameter of a pencil, two on those the diameter of your thumb, and two on those you're not sure of. If you do want to follow a specific system, the Four-Cane Kniffin is probably the best bet for the small grape grower.

However you prune them, your grapevines will need tying. Binding them up serves several functions. For one thing, the vines grow where you want them to, making for easier picking and covering your arbor instead of trailing along the ground. For another, tying minimizes wind damage, which could otherwise be severe. Lastly, it prevents the formation of contorted trunks that lessen the yield. Don't use wire to bind the vines to their supports. It may look perfect, but it will cut into the bark and girdle the arms. Jute twine or some other soft string is good. Two-ply or stronger is best; it will cut less. Wrap the twine several times around the arbor poles or cross-wires, then prop a cane up against it and

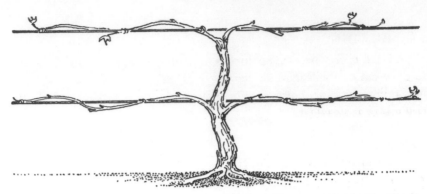

GRAPEVINE PRUNED BY THE FOUR CANE KNIFFIN
(SINGLE TRUNK) SYSTEM

knot it, loosely but so it cannot untie—the old granny knot is fine. Be careful not to break the canes when tying them, don't use more ties than necessary, and space out growing canes evenly.

It's the second winter's pruning that begins to shape up your vines. If you're using the Four-Cane Kniffin system, the first year's growth is cut back in winter to within a few buds of the initial pruning. The second year you will need a stake for the vine to climb up if you don't have the arbor ready yet. Let the plant leaf out fully to insure that it retains its health. Wait and winter-prune, again leaving a little more wood than the vine had before. Each of the main canes (the big fat ones) should be left with ten buds. All laterals except one should be cut off each cane. These laterals will become main canes the following year. Train the laterals to keep growing along the horizontal. The new spring growth the third year can then be trained vertically to completely cover the arbor. Your best bunches of grapes will be found around the fifth to eighth nodes, or leaf junctures, of the new growth. Anything beyond that can be cut off.

PROBLEMS

One of the most common problems the small vineyard owner faces is "mummy grapes," or black rot. The grapes decide to try and become raisins on the vine. But it doesn't work; they rot from the inside out, acquiring a wizened, shriveled look. This fungus disease is best controlled by immediate removal of the offending bunch before the fungi can spread. You can't cure it, you can only minimize the effects.

Japanese beetle was once a frequent arbor plague. It can be controlled by introducing milky spore disease, which leaves the grapes unaffected but wipes out the beetle.

Healthy rootstock and the precautions mentioned above should take care of any problems you might have. Grapes as a rule are not troubled by many diseases and pests.

HARVESTING

Let your grapes ripen on the vine till stems begin to dry slightly, the aroma is superb, and they are at their sweetest. Since you're not using all sorts of chemical sprays and fertilizers, you can sample them right off the vine.

Pick grapes in the cool of the morning. Clip the bunches off with scissors rather than trying to twist the stems loose. Put the clusters in a shady spot, inside the arbor, for instance, as you harvest. Try not to eat more than three pounds while picking or you might be sorry.

STRAWBERRIES

Although they can be cultivated almost all over the United States, strawberries from cooler areas are the tastiest. Florida strawberries are popular among supermarket shoppers because they're available in the wintertime and they look good, but as to their taste, well, even if you live in Florida you'll be surprised how superior your home-grown fresh-from-the-garden ones are.

As to the varieties to choose from, even if you tried a different one for every plant in a hundred-foot row, you'd barely scratch the surface. Experiment to find out which ones suit you best, but remember to get some each of early, midseason, and late varieties to insure the longest possible harvest. Everbearing berries produce three crops, but, as with everything else when you try to make too much of a good thing, these crops are often second-rate. Besides, they take much more work watering, blossom-pinching, etc. Stick with a triumvirate of the regulars.

PROPAGATION

Strawberries can be grown from seed, but this is usually done only commercially in an attempt to increase even further the number of varieties available. For garden propagation, the plant is considerate enough to send out runners, which develop into new plants. In selecting plants, always buy ones certified virus- and disease-free, and you're more than halfway home with a good crop.

LOCATION

Strawberries need rich, light soil full of organic matter. A pH of 5.5 to 6 is good, since they like earth that's a bit acid. Proper drainage is necessary to minimize disease. Slope-planting is excellent; halfway up a northern hill will minimize the chance of flower destruction by a late frost, if this is a problem in your region. Remember to plant with the contour if the slope is at all steep. Full sun and placement away from trees, grape arbors, or other producers of competitive surface roots is a must, as is plenty of ground moisture, particularly during fruiting.

PLANTING

Very early spring, on a cloudy day, is the best time to plant strawberries. Start with a hundred plants and you'll have five to six hundred yielding ones the next season.

Prepare the land the previous fall by digging up a row to a depth of one foot. Incorporate as much compost and manure as you can, up to 50 percent of the total volume if possible. But remember, the compost should be acidic. Don't use lime unless the pH is well below 5.5. Although the roots will not often penetrate to the one-foot level, this depth of preparation aids in water retention. Also, since the soil is loose to this depth, the roots will grow deeper than they normally do, producing sturdier plants. Sow the bed to a winter cover crop of rye beforehand.

To minimize disease problems strawberry plants should not be set in soil that within the past two years grew tomatoes, potatoes, peppers, eggplant, raspberries, or corn. Taking this precaution will minimize the chances of both verticillium wilt and root lice.

Plant at intervals of one and a half feet with four feet between rows. To take, the roots should be quite moist when planted. If they look at all dry, soak them in a tray of water for half an hour before planting. While you work in the field, keep all the plants covered with wet cloth until you set each one, to insure that they don't dry out. The crown must be set into the soil just as deep as it was before. Too high or too low a level will mean the end of your plant. The roots should be evenly spread out and the soil on top of them packed down quite firmly.

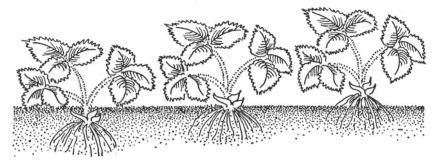

SET TOO LOW   PROPER SETTING   SET TOO HIGH

SETTING STRAWBERRY PLANTS

TO GIVE THE PLANTS THEIR BEST START, MAKE SURE THE NEW SOIL LEVEL IS EXACTLY EVEN WITH THE CROWN.

A quick test of how you're doing with your planting is to go back to a plant you set a few minutes earlier, take hold of one leaf, and give it a quick, sharp tug. If the plant comes up, you didn't pack the soil enough. If the leaf snaps off its stem, you're all right, Jack.

Douse each plant heavily for its baptismal watering. If the weather seems particularly dry the day you're planting, snip off 10 to 25 percent of the lower leaves, in order to aid moisture retention by cutting down on leaf transpiration.

As strawberry plants grow, they send out runners that develop new plants. These can be trained in specific patterns for a larger yield. Simply anchor the runners with a bit of earth to keep them growing in the right direction. Planting more strawberries than you expect to harvest and just trimming off some of the runners in not too systematic a fashion also works fine. Just cut the runners once an old plant has borne its fruit. When an old plant stops bearing, dispose of it.

MULCHING

The strawberry was one of the first plants to benefit from the idea of mulching. Along with all its other beneficial effects, mulch prevents the berries from touching the ground, keeping them in better shape. Mulch yours with straw right after planting. If the mulch later prevents the runners from taking root, nudge the mulch aside beneath each new

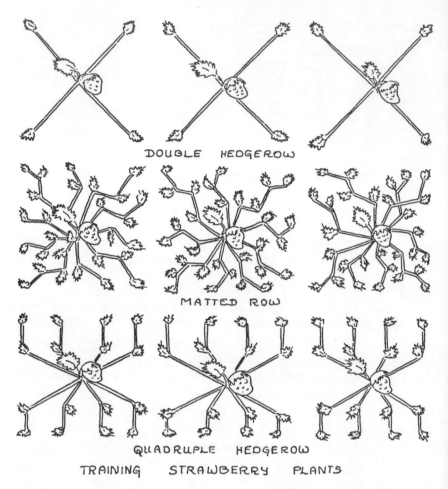

DOUBLE HEDGEROW

MATTED ROW

QUADRUPLE HEDGEROW

TRAINING STRAWBERRY PLANTS

PINCHING OFF UNWANTED RUNNERS AND ANCHORING DOWN REMAINING
ONES TO A SPECIFIC PATTERN MAKES FOR EASIER PICKING AND BIGGER YIELD.

runner and show it the way to the soil. If there is low snowfall in your
area and winter temperatures fall below 25°F. for any length of time,
mulch your plants for the winter with six to eight inches of straw after
the first frost, when the plants will be completely dormant. If you're assured of plenty of snow, it will work just as well for insulation. Remove
the straw blanket in spring when the days warm up and the strawberries
look as if they're about ready to grow again. That is to say, they will
have begun to leaf out, but the leaves, without sun, will be yellow. Just
gently move the mulch clear of each individual plant and let it lie.

PRUNING

Strawberry plants don't need pruning the way your other berries do. However, the first year they should have all their blossoms removed so they won't bear. This insures good rooting. The same procedure does not have to be followed with plants propagated from the runners, since they establish their root system the first year of growth, while still attached to the umbilical cord of the mother plant. In subsequent years pinch off most of the runners before they produce plants, leaving only enough to replace those that might die, or to permit a slight expansion of your strawberry patch. Otherwise you'll be flooded with plants. Also, pinching off excess runners will increase the yield of the parent plants.

In fall, just before mulching for winter, the new plants should be thinned or replanted so that there is a minimum space of twelve to sixteen inches between all plants. Fertilize liberally and apply the winter mulch.

HARVESTING

Like most berries, strawberries are best picked in the cool of the morning when they are at their peak flavor. Don't pull, twist them gently from the stem when they look ripe. Sample a couple, naturally. Some berries ripen with more white than others. Tasting is the sure test.

Vegetables

A plant is like a self-willed man,
out of whom we can obtain all which we desire,
if we will only treat him his own way.
—GOETHE

FOR THE SMALL FARMER, no single crop, area for area, will yield as much as the vegetable garden. When James Norman Hall was living in Tahiti, there was a shortage of familiar fruits and vegetables, and what few there were, were motley. The papayas and mangoes tasted great, but occasionally he longed for a good lettuce and tomato sandwich. Or so the story goes. He had a friend of his send him a large package of assorted superseeds to plant. Hall was busy with his writing, however; his garden suffered from neglect, and eventually expired from total insect devastation. He gave the remaining seeds to some Chinese friends, thinking that perhaps they could at least salvage a few vegetables for their family. Five months later, after carefully tending their new garden, his friends returned a huge basket of vegetables to him. Two years later he was still getting a basket a week—and the industrious farmers were supplying almost all the fresh vegetables to the town of Papeete. And that's more or less the story of any vegetable patch. Good seed and good care give superabundance.

How big you want your garden to be depends not only on how much vegetables are part of your diet, but which ones as well. The yield of tomatoes, say, is much greater than that for sweet corn occupying the same amount of ground. A one-hundred-by-one-hundred-foot garden should yield an ample supply for four plus guests. To help you plan your garden, the approximate yield per hundred-foot row is given for each vegetable in this chapter.

LOCATION

If you keep chickens, the likeliest spot for your vegetable garden is someplace where you can fence off half the area for a chicken yard and plant on the other half. The next year, rotate so that what was the chicken yard becomes the vegetable garden, and vice versa. The double yard makes for healthier chickens and gives you prefertilized soil to work with as well.

The garden should be on well-drained soil, preferably a gentle slope to the south if you have one. Sloping assures drainage, and southern hills warm up the best during cool springs and autumns, giving the plants on them extra growing strength. The difference between flat land and a sunny southern slope can mean as much as a two-months-longer growing season. But don't forget to follow the contour of the hill with your rows, or erosion will eventually wipe out the benefits of the site. Also, the tallest plants should be located at the northern end of the garden. A couple of rows of high sweet corn along the south side, for instance, could shade a large portion of your garden for most of the day. Your vegetables will want sun, every bit of it they can get, so if you can't give them a slope exposed to the south, make sure the garden is at least away from tall trees and in the open.

SOIL TESTING

Your field and forage crops will thrive without your knowing the soil too intimately. You'll know if it's acidic and needs lime; you'll fertilize the fields with animal and green manure. But since there you're working with large areas and less than absolute precision, if a couple of square feet in a field don't yield as heavily as the rest, your overall harvest won't be affected much. In the vegetable garden, however, if a hundred-foot row of potatoes doesn't grow because you've limed the soil, that's the end of your entire crop. Test the soil before you stake out and plant your vegetable garden, particularly those first few years when you're building up the soil and learning how things grow.

Most state agricultural colleges or experiment stations have facilities for running soil tests on your samples. But if you're not close to one or the other, the processing could take time and be a bit on the expensive side. If you're in a rush to get started, a soil-testing kit is a great time-saver. With it you can check things out right on the spot.

A simple pH kit is inexpensive and easy to use. You just moisten the innocent-looking paper against the soil. The paper changes color. Match that color with the chart, and presto, you get a pH number.

The larger, more expensive soil kits will give you, besides a pH rating and without much more work, a nitrogen-phosphorus-potassium breakdown. You put some soil into a test tube, add reagents, and match against another chart.

Remember when taking soil samples to use a clean shovel. If you use one that's been sitting in the manure pile all morning, you won't believe how rich your soil tests out—that is, you'd better not believe it. Also get samples from different parts of the garden, since soil conditions will vary. This doesn't mean you have to test each square yard and then supply the specific ingredients it lacks. Sample selectively the whole area you plan to cultivate, mix all the samples together thoroughly, and then test the mix. It will give you a good idea of what's needed overall.

FERTILIZER

The richness of your vegetable plot will depend on five factors, namely, nitrogen, phosphorus, potassium, organic matter, and trace elements. The basic fertilizer source for the garden will be the compost pile.

A good dose of compost, say two inches or more if you have enough to go around, should be dug into the garden two or three weeks before planting, while you're still tilling the soil and have committed yourself to what you're going to grow by leafing through all those glorious seed catalogs and finally cutting down the list of vegetables you'd like to manageable proportions before sending away for the seeds.

NITROGEN, PHOSPHORUS & POTASSIUM—WHAT THEY DO

Nitrogen: A basic building block of protein. Produces good stem and leaf development. Makes plants succulent and green.

Phosphorus :   Stimulates root development in young plants. Increases the proportion of fruit to plant. Speeds maturity of crops, also increases plants' resistance to some diseases.

Potassium :   Essential in the formation of starches, sugars, and cellulose. Without it plants do not mature well. Also aids in disease resistance, particularly those diseases caused by fungi.

If your soil test indicated a lack of nitrogen, phosphorus, or potassium, the deficiency can be remedied easily by adding material rich in the needed element to the soil. Apply it along with the compost before planting. If your soil is very low in one element, concoct a compost heap that will compensate for it. Low in nitrogen? Add extra animal manure, tankage, blood meal, fish scraps, and so forth. Phosphorus-deprived? Add phosphate rock, bone meal, tankage, incinerator ash. Low in potassium? Throw in some seaweed (wash the salt off if you gathered your own, otherwise it will build up in the soil), green sand, wood ash, or cocoa shell.

Once you've given your garden plot its initial composting, planted all those rows of good seeds and watched how they grow, more fertilizer may be added for the mature plants. Take well-composted material and spread it along the sides of the rows, being careful not to cover the growing plant stems. If you raise the soil level around the stem of a plant by stacking up compost against it, you may induce stem rot. Besides, the fine roots of the plant, the ones that absorb most of the nutrients from the soil, are some distance away from the main stem. This side dressing of compost between the rows will slowly filter down into the soil, providing continuous nourishment for those hungry little roots. At the same time it acts as a mulch, keeping down weeds and retaining moisture in the soil.

SEEDING

Many vegetables are given a head start by indoor planting. For those that take more than ninety days to reach maturity or bearing age, the early start is almost a necessity in the northern half of the country if you expect to get much of a harvest.

Plant the seeds in pots or wooden flats, or trays. If you use flats (preferred because they lend themselves to row planting), don't make them too long. You're going to be lugging them in and out of doors later.

Since the mature plants are going to have to adjust to life in the cold, cruel world outside, soil for the seedlings should not be superrich. Spread the bottom fourth of the flat with sphagnum moss or pure humus. Fill the rest of the flat with good-quality but not overly rich soil, leaving enough room to cover the seeds with an adequately thick layer. Sow the seeds thinly in rows, and over them spread compost to the proper depth for that type of seed. To give your seeds an easy start, use sifted compost as a cover. That is, take your best-quality compost and rub it through mosquito screening. Spray water gently over the top of the soil to help the seeds settle in.

Keep the beds moist by covering with glass or plastic, but lift the cover at least twice a day to circulate air. Water the trays by dunking the bottoms in water for a few minutes rather than watering from the top. This will aid root development while keeping the plants from damping off. Damping off is caused by a fungus disease that thrives when the surface soil around a young plant stem is damp. The fungus attacks the young, tender stem. The plant keels over and dies. So don't over-water. If any fungus develops on the soil, scrape it away immediately and leave the cover off. You'll be taking the cover off anyway when the first shoots break the surface of the soil. Don't let the plants crowd. If several have come up in one spot, pull out the weakest ones. Keep the flats indoors in a warm, sunny spot.

When the first two regular leaves, as opposed to the seed leaves, have developed, the seedlings should be spaced out, two inches apart, in two-inch rows. Dig them up with a soil bundle around the roots, plant them in a new flat filled with ordinary soil that is friable, but not too rich. If the soil seems to pack tightly, mix in some sharp sand or vermiculite.

Somewhat poor soil at this stage of the game will make the roots shoot around looking for better stuff. This gives your plants a solid foundation for future growth.

The new flats are still kept indoors in a sunny spot with night temperatures not above 60°F. However, as soon as the days begin to warm up, take them out to the yard daily, and bring them in again at night. This will acclimatize them to conditions outside. After a week of exercising yourself by carrying the flats back and forth, the plants are ready to be transplanted to the garden. Wait till after the last killing frost in your region, of course.

A cold frame can be used to house the flats outdoors instead of lugging them around. The principle of the cold frame is that it blocks off chilling winds, at the same time permitting the radiant heat of the sun to penetrate, keeping the plants all warm and cozy. The cold frame is opened during the warmest part of the day to prevent steamed vegetables and allow the circulation of fresh air. The degree to which it is opened depends on the day's temperature. Your objective is to keep the inside temperature in the 60° to 70°F. range. A thermometer will give you an indication of how much to open the frame. Close it early in the afternoon so enough heat is retained to keep the plants cozy at night. If you find night temperatures inside the cold frame falling below 40°F., you should move the seedlings indoors.

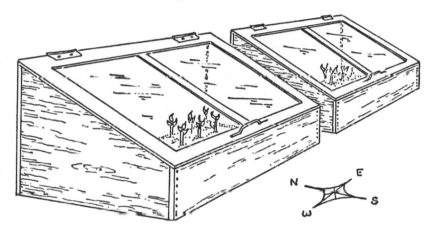

COLD FRAMES MADE FROM OLD BASEMENT WINDOWS

When growing seedlings in flats, indoors or in a cold frame, a week before they are to be set out in the garden take a sharp knife and cut the soil to the bottom of the flats so that each plant has its own individual cube of earth. This will cut any overlapping roots. At the same time, the plants will have a week to recover from the surgery and develop new hair roots within their own territory, thus lessening transplant shock. Individual peat starter pots save a lot of work and trouble, and with mass production their costs are declining to the point where they might be worth considering.

Transplant from flats to the garden when all danger of frost is past and the plants have developed numerous healthy leaves. For an extra head start, you can set out some plants before the frost-free date, but make sure to mulch them heavily. Late afternoon or evening is the best time to transplant. Water the garden well to insure settling in of the roots. Since young plants should not be exposed to direct sunlight the first two days after being transplanted outside, and since cutworms will try to do to your plants just what their name implies, it's a good idea to put a paper collar around each plant and cover the whole thing lightly with mulch. Don't mulch between the rows of collared plants, however, or the mulch will retain too much of that good dousing of water you gave the garden, and also keep the soil from warming up in the sun.

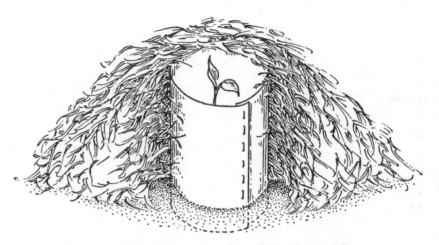

CUTWORM COLLAR WITH MULCH COVER

HEAVY PAPER COLLAR STAPLED TOGETHER PROTECTS YOUNG TRANSPLANT FROM CUTWORMS. COVER ENTIRELY WITH STRAW MULCH TILL ALL DANGER OF FROST IS PAST OR, IF THERE IS NO DANGER OF FROST, FOR TWO DAYS TO KEEP PLANT FROM GETTING TOO MUCH SUN WHILE IT SETTLES IN.

If you think you're going to have enough to do around the farm without cold-framing and transplanting a whole lot of seedlings, except maybe the tomato and pepper plants, you can still speed up getting the vegetable patch going. Cloche gardening—covering crops in the field with portable glass or plastic greenhouses—makes direct planting in the garden possible earlier. This type of temporary greenhouse has given excellent results in extending the growing season in the more northern latitudes, such as New England.

The cloche protects seeds and plants from frost. At the same time, moisture is retained, and rain running off the top and sides enters the ground away from the plants, assuring that the soil directly around them does not cake or pack, even without hoeing.

MODIFIED CLOCHE GARDENING USING PLASTIC SHEETING WEIGHTED WITH ROCK

A modern version of the cloche can be made by inverting U-shaped frames at one-foot intervals over the rows to be seeded. Coat hangers or willow branches work well. The maximum width you'll want at the base is twenty inches. After seeding, the upside-down U's are covered with polyethylene. Each end of the tunnel is folded over to retain heat and moisture. Open the tunnel around eleven in the morning for ventilation and close at three or four o'clock for the evening. But be sure to keep the tunnel closed during windy days or it will take off.

There's still, of course, the old tried and true method of planting a vegetable garden. As soon as the frost season is past, you sow the seeds in the garden, cover them up to the appropriate depth, using sifted compost if possible, and stand back. Except for weeding, weeding, and weeding.

But do you have to weed? Not any more. A well-kept garden on healthy soil is easy to care for if you use mulch.

MULCHING

By spreading a layer of organic material on the soil around your crops you can assure that the soil retains its moisture, and most weeds are eliminated because when they germinate and come up in the darkness beneath the mulch they wither away. Meanwhile, this mulch slowly decays, adding more natural organic material to the soil.

You can mulch along or between the rows of seedlings you set out from flats at once, but, except for mulch on top of a cutworm collar, it's better to wait a couple of days, since you've watered the plants heavily to insure root setting and you don't want the mulch to retain a lot of excess water. When you do mulch the young plants, leave an unmulched circle about six inches in diameter around each one for the first two weeks. After that you can bring the mulch almost up to the stems.

When sowing seeds directly in the garden, mulch between all the rows, but not right on top of where the young plants will come up. Wait till the plants are well formed, then weed around them if necessary, and mulch again. Mulch whenever the old material has decayed and compacted enough to permit weeds to sprout. You never have to weed an established garden. When you see some unwanted plants, just cover them with mulch. Not only will they die from lack of light, but they will contribute to the soil as they decay.

Mulch can vary from an inch to a foot in thickness. The basic rule of thumb is, the finer the material, the thinner the layer. When mulching with something like hay, you approach the one-foot level—after your plants are considerably taller than that, of course. Finer mulch, such as ground corncobs, peanut or cocoa hulls, or sawdust, need only be an inch or two. One last word: The soil must be able to breathe through the mulch. For this reason you can't use materials, such as leaves, that will bind together into flat cakey layers with no air penetration. At least don't use them directly above the plant roots; between the rows is all right if you absolutely have to. But then, you shouldn't have to, since they'd be better off in your compost pile.

So much for cultivating your garden. Just stand back and wait for the harvest. It will be the better for the mulch. Cucumbers, melons, and other ground-lying crops, for instance, won't become soft and rot, because they're resting on loose, dry mulch, not the ground. As for the harvesting, well, you'll know a country-fresh sun-ripened tomato or pepper when you see a real one, and a sweet ear of corn or pod of tiny new peas once you've tasted your first.

WHAT TO PLANT, AND WHEN AND HOW

You'll have no trouble making a grocery list for your garden. But the varieties to choose from for each vegetable often make almost as long a list. Those recommended here are good basic ones for the beginner. There might, however, be varieties better for your specific region. What are your neighbors growing? Try those, but don't forget to experiment a little too. You may find a couple of your favorites missing from the list that follows. Well, some of us never could stand Brussels sprouts. . . .

Planting periods for specific vegetables (given under the heading "Time to plant" in the charts that follow) vary greatly with the region you're in. The things to remember are: the average last killing frost is just that, the *average* date of the last killing frost in your region; the frost-free date is the date after which there will be no frost at all (not all frosts are severe enough to kill the plants), and this is usually two to three weeks after the average date of the last killing frost. So much for the frosts. "Early spring" refers to the week or two after the average last killing frost. "Very early spring" means you can begin planting even before all danger of frost is past.

Incidentally, I'm all for overproduction rather than under. In fact, continuous production is to your benefit. Sow seeds of any given vegetables once a week over the duration of their planting period. This will turn your garden into a movable salad bowl. And nothing goes to waste on a farm. Extra vegetables can always be frozen, canned, pickled, salted and stored in the root cellar for a winter supply or fed to the livestock. Even if they spoil, they're returned to the soil by way of the compost pile. Besides, if you find that five of your tomato plants are yielding all you can eat and you've got twenty, you can always take the extra tomatoes with you to town the next time you go and sell them to the greengrocer. You won't make a mint, but it'll pay for the gas. Or put up a roadside stand if you're well located for one. And of course you'll have friends dropping by who will be more than helpful in removing your excess yield.

ASPARAGUS

Variety:	Mary Washington
Hardiness:	Very hardy
Soil pH:	6.5
Time to plant:	Early spring
Seed longevity:	Roots should be planted on receipt
Seed/100-foot row:	75 roots
Distance between plants:	18 inches for rootstock
Distance between rows:	5 feet for rootstock
Depth to plant:	10 inches for rootstock

Seed that will germinate: 90% of roots minimum
Maturity: 2 years
Yield/100-foot row: 30 pounds

Asparagus takes more work initially than most garden vegetables. On the other hand, you'll be rewarded for your labors with a crop bearing for at least ten years, and in some cases up to fifteen or twenty years.

Remember to place your asparagus bed in a part of the garden that will remain untouched in the future, because it's going to be there awhile. Usually asparagus is planted at the edge of the garden, sometimes even completely separate from it in a corner of the yard—but not where it will be shaded by trees, and not where the chickens will be ranging every other year.

A deep, loamy, fertile soil that is well drained suits the plant best. The last requirement is the most important; asparagus will grow in almost any soil as long as it is well drained.

First thing, as soon as the ground can be worked in spring, decide how long you are going to make the bed. Two fifty-foot beds will yield as much as one a hundred feet long. Drive in a stake at either end of what will be the bed and draw a string between them.

Now comes the work. Dig a trench twelve inches deep by twelve inches wide the whole length of the bed. Over the bottom of each hundred feet of trench spread twenty pounds of ground limestone and fifteen pounds of phosphate rock. Moisten and dig into the bottom of the trench. Cover with a four-inch layer of quality compost. Some well-rotted manure, four months or older, may also be added. Work it all into the bottom again.

Lay your asparagus roots at eighteen-inch intervals along the trench; they should be about ten inches from the top. It's a good idea to buy 50 percent more rootstock than you intend to plant and dispose of the small weak ones. Cover with two inches of fine sifted compost. Water well and wait two or three weeks. Add another inch of compost and soil to the trench every two weeks until the trench is filled. This is only an approximate schedule. The object of the game is to let the tips keep growing a little higher than the soil level at all times.

By the time the first freeze of the fall approaches, the trench should have been filled. However, additional winter protection is needed. Mulch with compost, hay, well-rotted manure, or whatever is handy. So much for the first year's work.

Come spring, you rake off the mulch and leave it between the rows, or on the sides if you have only one row. Loosen the soil in the pit gently to a depth of six inches. No more, or you might injure the rootstock. Now make a hill of compost about one inch high over the length of the pit. As the warm spring weather continues, shoots will develop. Let them reach a height of six to eight inches; then, digging around them gently, harvest them by cutting off two inches below the soil. They must be picked daily or they go to seed. The first year's harvest shouldn't be extended for more than two weeks. The second year, cutting may be extended to a total of six weeks. This is the optimum length of time for the following years' harvest as well.

After your harvesting is done, shoots will keep coming up. Let these develop into lacy asparagus bushes, essential to the strength of your bed. Leave the bushes as part of the winter mulch.

Mary Washington is resistant to asparagus rust. Still, the beds should never be located in areas of frequent dampness, heavy dews, or mists, because these encourage the formation of rust.

BEANS, LIMA

Variety:	Fordhook
Hardiness:	Very sensitive, even to cold weather, much less frost
Soil pH:	6
Time to plant:	2 weeks after snap beans till early summer
Seed longevity:	3 years
Seed/100-foot row:	1 pound
Distance between plants:	10 inches
Distance between rows:	3 feet
Depth to plant:	1 inch
Seed that will germinate:	70%
Maturity:	75–90 days
Yield/100-foot row:	2 bushels

Once the plants begin to develop, say, when they're six inches tall, apply manure as side dressing. That is, place manure on either side of the plant, but no closer than two inches, since it can be burned by direct contact. This is particularly true of chicken manure; use it only for digging into the soil in the fall for next year's planting.

BEANS, SNAP

Varieties:	Bountiful (green), Stringless Green Pod, Golden Wax (yellow)
Hardiness:	Very frost-sensitive
Soil pH:	6.8
Time to plant:	Early spring to 60 days before average first killing frost
Seed longevity:	3 years
Seed/100-foot row:	1 pound
Distance between plants:	4 inches
Distance between rows:	2½ feet
Depth to plant:	½ to 1 inch
Seed that will germinate:	80%+
Maturity:	50–65 days
Yield/100-foot row:	2 bushels

Snap beans used to be called string beans, and still are in most places. But they shouldn't be, since the Department of Agriculture decided the old name made poor public relations and snap beans was a snappier image. They still taste great, whatever you call them, fresh off the vine, picked just before maturity. The plant forms a bush about eighteen inches high.

Sow the seeds directly into the garden. Cultivate very shallowly for weeds. Better yet, mulch away the weeds. The plants grow in almost any soil as long as it is not too acid. However, beans do best on rich soils, so compost well.

BEETS

Varieties:	Egyptian (early), Detroit Dark (late)
Hardiness:	Hardy, not injured by light frost
Soil pH:	6.5
Time to plant:	Very early spring to midsummer
Seed longevity:	4 years
Seed/100-foot row:	1 ounce

Distance between plants:	3 inches
Distance between rows:	1½ feet
Depth to plant:	½ inch
Seed that will germinate:	60%+
Maturity:	50–60 days
Yield/100-foot row:	1½ bushels

The beet is a native of North Africa, particularly around Egypt, from where the variety by that name came. Beets will thrive on just about any soil as long as it's not acid. They attract few pests and are in general very easy to grow. Since they have deep roots, loose, friable loam will give best results. If your soil tends to cake, incorporate plenty of organic matter. Don't use fresh manure. Well-rotted manure is fine.

When the tops reach the six-inch stage, beets often have to be thinned. This is because the "seed" is not an actual seed, but a fruit containing several seeds. The pulled beets can be transplanted. But why bother? You'll no doubt have enough without them. Instead, use the tender pulled beets whole and raw in a good country salad. Although it's fun to grow superbeets, for the tastiest eating pull the baby ones when they are less than an inch and a half in diameter.

Beet greens are an excellent source of vitamins A and C and iron. So learn to think of them as a salad crop as well as a root crop.

BROCCOLI

Variety:	Italian Green Sprouting
Hardiness:	Semihardy, withstands light freeze
Soil pH:	6.5
Time to plant:	Early spring to early summer
Seed longevity:	3 years
Seed/100-foot row:	With 10,000 seeds per ounce, the smallest packet will do great things
Distance between plants:	1½ feet
Distance between rows:	2½ feet
Depth to plant:	¼ inch
Seed that will germinate:	75%+
Maturity:	80 days
Yield/100-foot row:	45 heads

This plant belongs to the *Brassica*, one of the most diversified vegetable groups. Included in the clan are cabbage, Brussels sprouts, kohlrabi, and kale. It is highly nutritious and easy to grow. Start a spring crop in flats and transplant after the last frost. For a fall crop, plant directly in the garden in late May. Soil should be deeply dug. Drainage is more important than soil quality.

A broccoli plant will be about two feet tall when ready for cutting. Watch carefully that the buds do not open before you harvest. Cut off only the main head. The little buds on the remaining stem will now head out. You can get six to eight cuttings off one stem.

The plant is not prone to disease, but it does attract cabbage worms. Crop rotation is a vital preventative. Always grow the whole *Brassica* group in rotation, so no member of the family grows on soil where another one has been cultivated in the past two years.

CABBAGE

Varieties:	Bugner, Globe, Golden Acre (white); Mammoth Red Rock (red)
Hardiness:	Hardy
Soil pH:	6.5
Time to plant:	Early spring to midsummer
Seed longevity:	4 years
Seed/100-foot row:	¼ ounce
Distance between plants:	18–24 inches, more space for late varieties
Distance between rows:	2½ feet
Depth to plant:	½ inch
Seed that will germinate:	75%+
Maturity:	65–120 days
Yield/100-foot row:	45 heads

One of the most popular of the garden vegetables, and rightly so. You'll want to grow a fair amount because of its good storage qualities and to make sauerkraut. Soil moisture is important, so mulch well. As to the soil itself, cabbage isn't fussy, but it needs cool, humid weather for the best heading. Hence the late-maturing fall plants are preferred. Fall plants also make bigger cabbage heads.

Sow both early and late varieties in flats first and transplant as four-inch seedlings, or about six weeks after indoor planting. Liming will help keep down some pests, such as clubroot. Others—for instance, the cabbage worm, cabbage looper, and harlequin bug—should be picked off by hand when spotted.

CABBAGE, CHINESE

Varieties:	Michihli, Pe-Tsai, Wong Bok
Hardiness:	Fairly hardy
Soil pH:	6.5
Time to plant:	Early spring to late midsummer
Seed longevity:	3 years
Seed/100-foot row:	Another ultra-small seed, one package will do
Distance between plants:	16 inches
Distance between rows:	2½ feet
Depth to plant:	½ inch
Seed that will germinate:	75%+
Maturity:	70–90 days
Yield/100-foot row:	65 heads

A crop of increasing and well-deserved popularity. Stores well in the root cellar for up to two months. Needs cool weather, otherwise plants will develop a seed stalk instead of the head.

Sow directly into the garden. Thin when three inches tall and use the young pulled greens for salad. Remaining plants will need good evening watering during dry spells. Mulching helps; Chinese cabbage needs richly fertilized, moist soil.

CANTALOUPE

Varieties:	Early Knight, Hearts of Gold
Hardiness:	Very tender, can't handle any frost at all
Soil pH:	6.5
Time to plant:	After danger of spring frost, seedlings earlier indoors
Seed longevity:	5 years
Seed/100-foot row:	½ ounce
Distance between plants:	3 plants to a hill, 5 feet between hills

Distance between rows:	5 feet between hill-rows
Depth to plant:	½ inch
Seed that will germinate:	75%+
Maturity:	75–100 days
Yield/100-foot row:	50 melons

Cantaloupes are actually a variety of hard-rinded muskmelon. Their cultivation parallels that of cucumbers. Plenty of moisture, particularly during fruiting, and rich organic material are the key. In addition, sunny warmth is necessary. Cool temperatures or even continuously cloudy days will decrease yield and quality.

Cantaloupes, and for that matter most melons, are planted in small, raised mounds of earth, or hills. This produces bigger and better plants. You don't have to build a pyramid; a hill four to six inches high and a foot in diameter will do admirably. Mix in some aged manure when building the mounds.

Melons are best when field-ripened and picked just before eating. The way to tell a ripe one from its underripe neighbor is to push gently with the thumb at the joint where vine and melon merge. They should separate after even a light touch when the fruit is really ripe.

The cucumber beetle attacks cantaloupes too. Since you should already have marigolds growing by the cucumber vine (see the chapter on "Pest Control"), you might think it a bright idea to plant your melons there as well. Don't. Keep the two as far apart as possible, and plant a second batch of marigolds near the cantaloupes. Crop concentration just makes for greater temptation for the bugs.

CARROTS

Varieties:	Coreless, Imperator, Nantes
Hardiness:	Hardy, withstand light freeze
Soil pH:	6
Time to plant:	Early spring to midsummer
Seed longevity:	3 years
Seed/100-foot row:	½ ounce
Distance between plants:	2 inches
Distance between rows:	1 foot
Depth to plant:	½ inch

Seed that will germinate:	50%+
Maturity:	70–75 days
Yield/100-foot row:	2 bushels

Another easy-to-grow crop, with few insect enemies. Well-drained, organically rich soil should be used. Drop the seeds on top, then cover them with half an inch of finely screened compost. Tamp this down with a board to help eliminate air pockets. Carrot seeds, at 24,000 to an ounce, are very small and need all the help they can get in covering up snugly.

Often carrots and radishes are sown together and the quick-maturing radishes harvested when the carrots are first thinned. Thinning is essential for a good crop. And there's nothing like the delightful earthy carroty smell that surrounds you while you're at this task. You'll find the tiny young carrots make delicious eating. Carrots for winter storage are left in the ground till just after the first frost.

CAULIFLOWER

Varieties:	Early and Late Snowball
Hardiness:	Hardy
Soil pH:	6.5
Time to plant:	Early spring to early summer
Seed longevity:	4 years
Seed/100-foot row:	4 ounces will do for a whole acre, so get the smallest packet
Distance between plants:	1½ feet
Distance between rows:	2½ feet
Depth to plant:	¼ inch
Seed that will germinate:	75%+
Maturity:	90–95 days
Yield/100-foot row	40 heads

A tricky crop and not worth growing the first year unless you're a cauliflower nut. If you can, settle for broccoli. If not, sow in cold frames, and transplant seedlings after six weeks. Needs soil very high in organic matter, moisture, and nitrogen. Water daily every evening during dry spells.

Keep a sharp lookout for the young flower. When it is three inches in diameter, you have to tie the cabbage leaves over the flower like a Christmas present. Bows are not necessary, but the flower must be totally protected from light. Another method is simply to break the leaves over the flower—of course, one good heavy wind and goodby cauliflower crop. Cauliflower heads develop very rapidly. They'll be ready for cutting five to twelve days after bundling; the warmer the weather the faster the development.

CHICORY

Varieties:	Witloof (for French endive), Magdeburg (for coffee)
Hardiness:	Very hardy
Soil pH:	6
Time to plant:	Early summer
Seed longevity:	4 years
Seed/100-foot row:	1¼ ounces
Distance between plants:	3 inches
Distance between rows:	2 feet
Depth to plant:	¼ inch
Seed that will germinate:	60%+
Maturity:	95 days
Yield/100-foot row:	2½ bushels

An all-around, handy, easy-to-grow plant not often found whole in the city. Young shoots, frequently sold as endive in city markets, are for munching, the roots for boiling like carrots or turnips. What makes chicory particularly valuable is that it can be forced in winter to produce fresh salad greens. To boot, the dried, roasted roots when ground make a good coffee substitute, and for all your farm's self-sufficiency, you're not going to be able to grow real coffee in a temperate climate. Nor real tea, for that matter.

It is a hardy crop, but likes well-drained, heavily manured soil. Hardly any insect enemies or disease problems.

For winter endive, leave the roots in the ground through the first few frosts. Dig them up and place them upright in a box of soil with about two or three inches of soil between the plants. Cover each crown with loose, moist sand to a depth of seven inches. Store in an area where they will not freeze, but where the temperature does not go much above 40°F. Keep moist with a weekly watering. Two to three weeks before you want to use them for salad, place the box in a room where the temperature is 60° to 65°F. Give the spent rootstock to your animals for a bit of winter variety.

CHIVES

Varieties:	No separate varieties
Hardiness:	Hardy
Soil pH:	6
Time to plant:	Early spring
Seed longevity:	2 years
Seed/100-foot row:	½ ounce
Distance between plants:	1 foot
Distance between rows:	2 feet
Depth to plant:	¼ inch
Seed that will germinate:	60%+
Maturity:	85 days
Yield/100-foot row:	100 clumps (or more than enough for a family of thirty-two)

Another handy plant for the homestead in need of winter greens. Take a couple of the clumps indoors, planted to trays, before freezing weather sets in. They'll give you tasty fresh sprigs for soups all winter.

Chives like a very sunny location and rich, loose, pebbly soil. The second spring, clumps left in the garden over the winter can be dug up and the bulbs at the bottom of each bunch divided so you get two plants. Cutting for salad should be done with gusto, not trepidation, for the more you cut, the better the plant survives.

CORN, SWEET

Varieties:	There are probably five billion varieties by now, but the old Golden Cross Bantam is a good way to settle the confusion
Hardiness:	Tender
Soil pH:	6.5

Time to plant:	Early spring to midsummer
Seed longevity:	3 years
Seed/100-foot row:	¼ pound
Distance between plants:	15 inches
Distance between rows:	2½ feet
Depth to plant:	1 inch
Seed that will germinate:	75%+
Maturity:	80–90 days
Yield/100-foot row:	50 ears

Yes, the Indians were right, a dead fish in each hill of corn does help. But before you raid your sardine tins, in case you're not near one of the coasts, try instead a good application of well-rotted manure two weeks before planting, or fresh manure the preceding fall.

Corn is hard on the soil, and fertilizer, particularly one rich in nitrogen, should be used liberally. A winter crop of clover is a good idea for this part of the garden. Mulching and side-dressing are worthwhile.

Sweet corn should be harvested in the milk stage—and, of course, the last possible minute before dinner. Break a kernel with your fingernail; if the inside is milky liquid, it's ready. Harvesting early, before the kernels reach the dough stage, is particularly important if there are many crows in your area and you'd still like to eat some of your own corn.

Save a little dent corn from the field crop, if you plant one, for parched corn, a tasty change from popcorn on a winter's evening.

CUCUMBERS

Varieties:	Early Fortune, Markater; for pickles, Black Diamond, West Indian Gherkin
Hardiness:	Very tender, tolerate no frost
Soil pH:	7
Time to plant:	Spring, after all danger of frost is past
Seed longevity:	5 years
Seed/100-foot row:	½ ounce
Distance between plants:	3 plants to a hill, 5 feet between hills
Distance between rows:	6 feet between hill-rows
Depth to plant:	½ inch

Seed that will germinate: 80%+
Maturity: 70–75 days
Yield/100-foot row: 1½ bushels

Will not grow in acid soils. Need loose, open soil. Sandy soils are good if organic matter is high. Plowing under animal manure the fall before planting is very helpful.

Unless mulched, cucumbers will need thorough weekly cultivation. Plant six seeds to a hill and thin out the three weakest when the plants are four inches high. Making one-foot-wide, six-inch-deep trenches between the cucumber rows and filling them with well-rotted manure covered by an inch of soil and some additional mulch helps assure a good crop in dry weather. Water the trenches daily. The rich seepage will boost plant growth.

Six to seven weeks after planting, small cucumbers will form. Watch them carefully, because they're like balloons being blown up. One day there's a mini-cucumber, the next it's fifteen inches long. Cucumbers are over 96 percent water. That's what fills them up so fast, and that's why the crop is so moisture-dependent.

Healthy cucumber plants shouldn't be affected seriously by pests or disease. However, the cucumber beetle, which also spreads bacterial wilt, must be kept in check. Plant marigolds around your cucumber plants. Personally, I never did like them, but neither do the cucumber beetles. And I'd rather have flowers than beetles. Wood ashes sprinkled on the soil as fertilizer are an additional deterrent.

GARLIC

Varieties: No special varieties
Hardiness: Hardy
Soil pH: 6
Time to plant: Very early spring
Seed longevity: 1 year for cloves
Seed/100-foot row: 2 pounds (300 cloves)
Distance between plants: 4 inches
Distance between rows: 2 feet
Depth to plant: 2 inches
Seed that will germinate: 90%+

| Maturity: | 90 days from clove sets |
| Yield/100-foot row: | 10 pounds, which is enough to take your breath away |

Bulbs give better results than seeds, and results they do give. Don't drop any on your way to the garden or you'll have a Hansel and Gretel trail growing all the way back to the house come summer. Break the big bulb into cloves, or sets. These sets, composed of two or three cloves each, are what you then sow. Nitrogen fertilizer is a plus, although if you have good soil, nothing will be needed.

Harvest in midsummer or when the tops ripen, dry, and begin to fall to the ground. Pull up the whole plant. Let dry thoroughly, then braid the grass top parts together to form a long chain. Hang in a dry spot in the kitchen for ready use.

HORSERADISH

Variety:	Maliner Kren
Hardiness:	Hard to kill even with a blowtorch
Soil pH:	7
Time to plant:	Very early spring
Seed longevity:	Over the winter
Seed/100-foot row:	70 roots
Distance between plants:	1½ feet
Distance between rows:	3 feet
Depth to plant:	4 inches for roots
Seed that will germinate:	90%+
Maturity:	Harvest any time beginning 1 week after first leaves develop
Yield/100-foot row:	30 pounds

As a perennial, this is another one to plant where it can grow undisturbed for several years. Although horseradish will grow anywhere, to produce good straight roots it should have deep, loose, organic soil, with manure dug down to a foot. The sets that are sown for horseradish are the small roots attached to the main one. Lay these horizontally in a six-inch-deep trench, with the thicker end of the root four inches higher than the thin end—in other words, average depth four inches. You can leave it in the ground over the winter. In spring, just dig up and break off fresh sets and plant them. Tastiest in spring. Always great for your digestion.

LETTUCE

Varieties:	Bibb, Grand Rapids, Dark Green, Romaine, White Boston
Hardiness:	Hardy
Soil pH:	6.5
Time to plant:	Very early spring through midsummer
Seed longevity:	6 years
Seed/100-foot row:	Small package
Distance between plants:	16 inches
Distance between rows:	1½ feet
Depth to plant:	¼ inch
Seed that will germinate:	80%+
Maturity:	45–85 days
Yield/100-foot row:	80 heads

If you're used to eating iceberg lettuce, consider switching. The loose-headed lettuces such as romaine and those in the butter group are much easier to grow, and, besides, their green leaves are more nutritious than their blanched white counterparts. Like most garden vegetables, lettuce prefers well-drained, rich soil, and, like cucumbers, it needs plenty of water during the growing season. In fact, the two keys to a good lettuce crop are plenty of water and plenty of growing space.

Thin ruthlessly, for if the lettuce doesn't have the room to grow quickly, it won't grow at all. Thinned plants when in the four-leaf stage are readily transplantable to another row, so there's no waste involved. This transplantability makes the crop ideal for starting in flats before you can work the garden. But remember, lettuce prefers its weather orbit on the cool side; flats, indoors or out, should be kept at between 50° and 60°F.

Leaf lettuce will refill your salad bowl with fresh-picked greens many times over if you harvest only the lower leaves. The center ones will grow and grow. But pinch the top off the center to keep it from flowering.

ONIONS

Varieties:	Early Yellow Globe, Sweet Spanish (from seeds); Ebenezer, Italian Red (from sets)

Hardiness:	Hardy
Soil pH:	6
Time to plant:	Very early spring; seed may be sown when frost danger is not quite past
Seed longevity:	1 year
Seed/100-foot row:	½ ounce seeds or 2 pounds sets
Distance between plants:	Seeds at 2–3 inches or sets at 4 inches
Distance between rows:	1½ feet
Depth to plant:	½ inch
Seed that will germinate:	70%+
Maturity:	120 days from seeds, 100 days from sets
Yield/100-foot row:	1½ bushels (55 pounds/bushel)

An easy-to-grow crop if, not to sound repetitious, your soil is fertile and well drained. Can be sown from either seeds or sets. Sets are a no-muss, no-fuss method, since you don't have to start them indoors or in cold frames. Even if you have the space, it's usually better to start tomatoes, peppers, lettuce, and the like indoors or in the frames, taking the risk of sowing onion seeds directly into the ground, even before the last frost. If using sets, they should be no bigger than half an inch in diameter. Although logic might dictate the bigger the better, large ones are more apt to put out seed stalks, making the onion thick-necked.

Set onions are not as good for storage as seed onions, and young onions shipped up from the South for planting are the best of all for later storage. The reason young plants are often bought from the South can be traced to the light-sensitivity of the onion. While the plant will grow well enough with less than ten hours of sunlight, most varieties of onion won't form the onion bulb unless the plant gets twelve to fourteen hours of daylight. You need the onion bulb, the plant doesn't. It can always propagate through seeds. To get onions, it's essential that your plants be in the bulb-forming stage by early summer.

At maturity, when about three-fourths of the leaves have dropped to the ground, give them a helping hand and knock down the remaining ones with the backside of a rake. Two days later, pull the onions to the surface and leave them there to sun-dry for a couple of days. This will help cure as well as dry them. Now clip off the old tops to within an inch of the bulb. Place the onions in a single layer on open slatted racks indoors or outdoors under a rain cover of some sort—a porch or even just a sheet of tin will do. They need continuous air circulation around them. When the weather stays cool all day long, bag the onions in netting and hang them in a cool, dry cellar.

PARSLEY

Varieties:	Moss Curled (for leaves), Hamburg (for roots)
Hardiness:	Hardy
Soil pH:	6
Time to plant:	Early spring
Seed longevity:	1 year
Seed/100-foot row:	¼ ounce
Distance between plants:	8 inches for leaf, 2 inches for root
Distance between rows:	16 inches
Depth to plant:	⅓ inch
Seed that will germinate:	60%+
Maturity:	90–95 days
Yield/100-foot row:	90 bunches

Plant some of both varieties of parsley, that grown for its top as well as that, like the Hamburg, whose carrot-shaped root is a great ingredient in soups and *real* stews. Both varieties will survive the winter in the garden, even in most northern areas, if well mulched.

Parsley seeds germinate unenthusiastically, and it's suggested they be soaked in tepid water for twenty-four hours before planting to assist the process. Even so, they will take a month to germinate. Parsley likes a lot of nitrogen. Apply some compost rich in that element as side dressing to the plants as they grow.

Trim leaf parsley for your table from around the outside of the plant, leaving the center leaves to mature further. Root parsley is harvested in fall. It stores well in your cold cellar.

PARSNIP

Variety:	Improved Hollow Crown
Hardiness:	So hardy you can harvest it with an ice pick

Soil pH:	7
Time to plant:	Early spring
Seed longevity:	1 year
Seed/100-foot row:	½ ounce
Distance between plants:	3 inches
Distance between rows:	1½ feet
Depth to plant:	½ inch
Seed that will germinate:	60%+
Maturity:	120 days
Yield/100-foot row:	2 bushels (50 pounds/bushel)

A good crop for the North. Parsnip roots may be left in the ground through the whole winter and harvested as needed. Freezing improves flavor.

Soak seeds like parsley; plant in well-drained, light, loose soil that has been richly composted. The parsnip root goes down a foot and a half, so make sure the ground is loose to that depth. Mulch over the planting to keep it cool and moist till the first shoots break ground, then clear enough space for them to reach the sky.

PEAS

Varieties:	Dwarf Alderman, Lincoln, Melting Sugar (edible pod)
Hardiness:	Hardy
Soil pH:	7
Time to plant:	Very early spring, just before last frost
Seed longevity:	3 years
Seed/100-foot row:	1 pound
Distance between plants:	1–2 inches
Distance between rows:	3 feet
Depth to plant:	1–3 inches; the lighter the soil, the deeper the seed
Seed that will germinate:	80%+
Maturity:	65–80 days
Yield/100-foot row:	1 bushel

Cool weather is a must for peas. This is why early spring planting is essential. Mulching the young plants throughout the spring helps keep the soil cooler and gives good growth.

Peas belong to the legume family and make a good rotation crop for your sweet corn, which is convenient since both crops do best when not planted to follow themselves in the same spot. For your peas, manure in the preceding fall if you can.

To give peas a head start in the garden, soak the seeds in water, half covered, until they sprout. It will take them a few days, but when they do sprout, they grow quickly, so plant immediately. Fertilize the plants well when they are one and a half inches tall.

Be sure to grow some edible pod varieties, the kind you usually get in the city only in Chinese restaurants as snow peas. Harvest them just as the seeds begin to form. Delicious served whole and raw in salads, or barely cooked.

PEPPERS

Varieties:	Ruby King, the Wonders, World Beater (sweet) ; Long Red Cayenne, Tabasco (hot)
Hardiness:	Very tender
Soil pH:	6.5
Time to plant:	Plant in cold frame or indoor flats 5–8 weeks before last frost
Seed longevity:	2 years
Seed/100-foot row:	Small package
Distance between plants:	2 feet
Distance between rows:	3 feet
Depth to plant:	½ inch
Seed that will germinate:	55%+
Maturity:	115 days
Yield/100-foot row:	4 bushels

Need well-drained soil, but not too much nitrogen. Nitrogen makes the plant grow well, so well, in fact, it forgets all about making peppers.

Grow peppers at the far end of the garden from the potatoes, since potato bugs like peppers for a change of menu. When starting peppers in flats, keep the soil very moist until the seed germinates. However, once the young shoots develop, be a little more sparing. They need the moisture but are also prone to damping off. Make sure there's no chance of frost before setting them out in your garden. Mulching is an excellent way to retain ground moisture. But, as with tomatoes, wait till after the first month's growth to mulch, or the ground will also retain the cold, which makes the plants fruit later.

Harvest the fruit by cutting the stem one inch above it. Do not twist loose. Sweet peppers are usually picked green for shipping. On the farm, picking them when fully ripe and reddish adds flavor. Hot varieties such as cayenne and tabasco must be picked red.

POTATOES, SWEET

Variety:	Yellow Jersey
Hardiness:	Very tender
Soil pH:	5.5
Time to plant:	2 weeks after last frost
Seed/100-foot row:	Buy plants, use 80
Distance between plants:	16 inches
Distance between rows:	2½ feet
Depth to plant:	Plant slightly lower than previous level
Seed that will germinate:	95% of plants survive
Maturity:	Can only tell if ready to harvest by digging up a sample; begin checking in early fall
Yield/100-foot row:	1–2 bushels

Sandy loam *not* too rich in organic matter is preferred. A good poor-soil crop. Use plants instead of seed potatoes because unless you're guaranteed the potatoes haven't been sprayed with a growth inhibitor, you're not going to get anywhere with them. Also the plants will be more disease-resistant.

They like dry, hot weather. Don't mulch, but cultivate until the vines get big enough to shade out weeds. Harvest before the first frost of fall. Dig the potatoes up carefully. Bruising them will reduce their storage quality.

Sweet potatoes must be cured before storing. But curing is simple enough if you're in the South. Just spread them out on some dry hay on the ground and let them lie in the sun for the better part of a week. Turn them at least once. In cooler climes, keep them in a 90 percent humidity environment at 80° to 90°F. for ten days. Store them packed lightly in straw, at temperatures not below 50°F., with good ventilation.

POTATOES, WHITE

Varieties:	Cherokee, Green Mountain, Irish Cobbler
Hardiness:	Semihardy
Soil pH:	5.5
Time to plant:	Early spring
Seed longevity:	Over the winter
Seed/100-foot row:	½ peck cut seed potatoes
Distance between plants:	1 foot
Distance between rows:	2 feet
Depth to plant:	4 inches
Seed that will germinate:	95%+
Maturity:	100–120 days
Yield/100-foot row:	2–3 bushels (60 pounds/bushel)

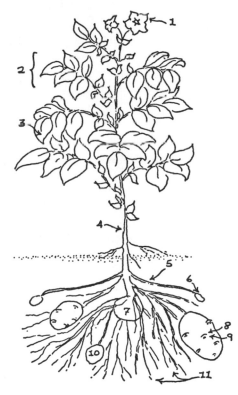

POTATO PLANT

1. FLOWER
   (BERRY POISONOUS)

2. LEAF

3. TERMINAL LEAFLET

4. STEM

5. UNDERGROUND STEM

6. INCIPIENT TUBER

7. OLD POTATO (SEED)

8. EYEBROW

9. EYE

10. YOUNG TUBER

11. TRUE ROOTS

Potatoes need acid soil, as opposed to most of your garden vegetables, which like things close to neutral. Don't put them in limed soil, soil that has been manured within a year, or that on which tomatoes have been grown for the past two years. Early varieties that get a cool growing start are less disease-prone. Potash is important, so use compost rich in it. Let cut seed potatoes dry for twenty-four hours before planting. Make sure each piece contains at least one eye for a sprout. Mulch potatoes with hay to one foot after planting; this will keep the soil cool for healthier growth.

Potatoes are ready for harvesting when the tops have mostly withered. Leave storage potatoes a little longer. Of course, don't forget to harvest small new potatoes early; there's nothing like them. Cure storage potatoes like sweet potatoes. Make certain they are dry before putting them away in a cool, dark place for the winter.

PUMPKINS

Varieties:	Cheese, Small Sugar
Hardiness:	Tender
Soil pH:	7
Time to plant:	Spring to midsummer
Seed longevity:	4 years
Seed/100-foot row:	½ ounce
Distance between plants:	5 feet between hills
Distance between rows:	5 feet between hill-rows
Depth to plant:	1 inch
Seed that will germinate:	70%+
Maturity:	65–75 days
Yield/100-foot row:	40 pumpkins

Pumpkins are often sown in rows between corn plants, as a double crop. The corn matures first and is harvested, leaving the field to the pumpkins.

Sandy loam is the preferred soil, although, again, this crop will grow in any well-drained, fertile soil that does not pack tightly. Usually planted six to a hill, with the weakest three or four seedlings being pulled as the plants begin to vine. It's a good idea to cover the hill and seedlings with a wooden or metal frame hung with cheesecloth to keep striped beetles away from the young, tender plants. The veil is removed, of course, once the plants begin to outgrow it.

If the vines seem to grow on endlessly with no flowers in sight, pinch back the tips of the vines. They'll take the hint. In the cool weather of fall, before the frost is on the pumpkin, cut the fruit from the vine, leaving several inches of stem. Let them field-cure for ten days and they're ready to be stored in your cellar till the big pumpkin-pie bake.

RADISHES

Varieties:	Cherry Belle, White Icicle, White Strasbourg
Hardiness:	Hardy
Soil pH:	7
Time to plant:	Early spring to end of summer
Seed longevity:	4 years
Seed/100-foot row:	½ ounce
Distance between plants:	Shoulder-to-shoulder for the red roots, twice that for the white
Distance between rows:	½ foot
Depth to plant:	½ inch
Seed that will germinate:	75%+
Maturity:	25–60 days
Yield/100-foot row:	100 bunches

Radishes will likely be your first crop, simply because you can sow them early and they'll go from seed to picking in just about a month. The red ones are faster-maturing than the white, but the white are larger and more flavorsome.

Loose organic soil will give you better, bigger roots. Don't plant spring varieties in the summer; they will yield poorly or go to seed. Other than that, the crop is about as easy to grow as they come, and the only care it needs is occasional thinning, which is accomplished by harvesting. Sow some more radishes every week for a continuous supply.

A frequently used trick is to mix radish seeds with beets, carrots, or other slow-germinating seeds. The emerging radishes indicate where the other crop will eventually come up so you don't hoe it away if you weed. At the same time, you get a bonus crop without extra ground preparation or seeding by pulling the quick-maturing radishes before the other plants begin their main growth.

RHUBARB

Variety:	MacDonald
Hardiness:	Hardy
Soil pH:	6
Time to plant:	Very early spring
Seed longevity:	Plant roots immediately upon receipt
Seed/100-foot row:	25 roots
Distance between plants:	3 feet
Distance between rows:	4 feet
Depth to plant:	2 inches
Seed that will germinate:	95%+
Maturity:	1 year to first picking
Yield/100-foot row:	200–250 stems

Rhubarb is a very healthy crop, particularly in northern countries where its vitamin C can be put to good use in the absence of citrus fruits. It can also be a deadly crop, however. Do not eat rhubarb leaves, nor feed them to your livestock. They contain large quantities of calcium oxalate, which is poisonous. Throw the leaves on your compost pile.

Rhubarb is another crop to plant off by itself where you and the chickens will leave it alone, not only to prevent leaf-nibbling, but because the plants will keep sprouting for years. Initial propagation by rootstock rather than seed is much preferred. Needs heavily fertilized, moist soil. Cool to cold winters suit the plant, although it should, under these circumstances, be mulched to a depth of four inches. Livestock bedding makes an excellent mulch, since the plant will use the manure. You can speed up spring growth by putting eighteen-inch-square open-top boxes around the plants so they have to grow tall quickly. Cut off any flowering stems that develop, as they decrease the yield of edible stems.

Six- or seven-year-old plants will have enlarged roots. This is the time to retrench your rhubarb bed. Dig up those plants that have given you the tastiest, most colorful stems. Cut the roots into five to eight sections, making sure that at least one bud remains on each root section. Plant the way you started out.

RUTABAGA

Varieties:	Improved Long Island, Perfection Swede
Hardiness:	Hardy
Soil pH:	6.5
Time to plant:	90–100 days before first freeze
Seed longevity:	4 years
Seed/100-foot row:	¼ ounce, or 2 pounds/acre drilled, 4 pounds/acre broadcast
Distance between plants:	4–5 inches or broadcast
Distance between rows:	1½ feet
Depth to plant:	⅓ inch
Seed that will germinate:	75%+
Maturity:	90 days
Yield/100-foot row:	2 bushels

An excellent storage crop. You'll want to grow enough not only for your own larder, but for your livestock's winter supply as well. Requires deep, moist, fertile soil, and prefers it cool, so it's usually sown as a fall crop.

If you're using rutabaga for stock feed, sow theirs a little earlier to increase the root size. It won't be quite as tasty, but still quite nutritious and satisfactory.

SPINACH

Varieties:	Bloomsdale Long Standing Savoy, Virginia Savoy (for late crop)
Hardiness:	Hardy
Soil pH:	6.5 (fairly pH-sensitive, best at this precise level)
Time to plant:	Very early spring to early fall
Seed longevity:	3 years
Seed/100-foot row:	½ ounce
Distance between plants:	2–4 inches

Distance between rows: 1 foot
Depth to plant: ½ inch
Seed that will germinate: 60%+
Maturity: 40–70 days
Yield/100-foot row: 3 bushels

Extra moisture and nitrogen-rich compost during the early stages of growth encourage an abundance of leaves. Keep sowing at two-week intervals even into the fall. Half-grown spinach is mulched heavily—up to a foot, depending on how far north you are. Remove the mulch in spring for early harvest.

For late planting, switch to a yellow disease-resistant variety such as Virginia Savoy. For areas with hot summers, plant New Zealand, which is not a true spinach but takes the heat better and cooks up just the same—if you must have your spinach cooked in spite of the crisp, colorful, vitamin-rich touch it adds to a garden salad.

SQUASH

Varieties: Straightneck (summer); Boston Marrow, Hubbard, Butternut (winter)
Hardiness: Tender
Soil pH: 6.5
Time to plant: Early spring to early summer
Seed longevity: 4 years
Seed/100-foot row: ½–1 ounce
Distance between plants: 5 feet between hills
Distance between rows: 6 feet between hill-rows
Depth to plant: ½ inch
Seed that will germinate: 75%+
Maturity: 55 days (for summer varieties), 150 days (for winter varieties)
Yield/100-foot row: 100–150 bushels

Plant like pumpkins, using extra manure in the hills. Winter squash—you can't store summer squash—must be harvested before frost. If you are growing gourd squash as well, let a few of the smaller ones dry out by cutting a hole in one side and setting them by your stove or fireplace. When thoroughly dried, shake out the seeds, and hang the gourds up by their stems from a tree branch, using short strings so they won't swing too much, or nail them to the tree. Make fine guest-houses for the bug-eating wren-sized birds that might visit your spread.

TOMATOES

Varieties:	Cherry, Beefsteak, Rutgers
Hardiness:	Tender
Soil pH:	4
Time to plant:	Start in flats in February or March; transfer to garden at 8 weeks
Seed longevity:	4 years
Seed/100-foot row:	Small packet
Distance between plants:	3 feet
Distance between rows:	4 feet
Depth to plant:	½ inch
Seed that will germinate:	75%+
Maturity:	115 days
Yield/100-foot row:	4 bushels

Flat-raised seedlings should be transplanted after they have developed three or four true leaves. Separate seedling containers are preferable, although seeds sprouted together can always be spaced out later in a fresh flat if necessary. Water the flats from the bottom by setting them in water for three to four minutes. Keeping the surface of the soil dry will minimize damping off in the young plants.

Fertile, well-drained soil with plenty of organic material is needed by tomato plants. For very heavy clay soils, add sand. The more sun the better. Tomatoes also need good air circulation around them. Mulch is excellent, but wait till the plants have a good sturdy start or the mulch, retaining cold as well as moisture, will stunt their fruiting. Pest control is more important for tomatoes than for almost any other of your garden crops, since everybody likes to eat tomatoes. And, unlike your other plants, tomatoes should be set one to three inches deeper when transplanted into the garden. New roots will grow on the buried stems, boosting the yield.

Tomato plants bear so heavily that you will have to stake them. If not supported, they will bear less, and the smaller fruit will often spoil. Allow only one main stalk to keep growing on staked plants.

Snap off all others while still small. Green tomatoes remaining on the vine by fall should be picked just before the first killing frost, either for pickling or for keeping on a warm sunny windowsill to ripen.

TURNIPS

Varieties:	Golden Ball Yellow, Purple Top
Hardiness:	Hardy
Soil pH:	7
Time to plant:	Spring to fall
Seed longevity:	4 years
Seed/100-foot row:	¼ ounce, or 1½ pounds/acre broadcast
Distance between plants:	2–3 inches or broadcast
Distance between rows:	1½ feet
Depth to plant:	⅛ inch, covering lightly is important
Seed that will germinate:	80%+
Maturity:	60–70 days
Yield/100-foot row:	2 bushels (55 pounds/bushel)

Grow the same way you do your rutabaga. Again, plant some extra for your livestock. Turnips make good storage fodder for the winter.

WATERMELON

Varieties:	Blue Ribbon, Dixie Queen, Sugar Baby, Midget
Hardiness:	Very tender
Soil pH:	6
Time to plant:	Early spring, started in flats in the North
Seed longevity:	2 years
Seed/100-foot row:	½ ounce
Distance between plants:	8 feet between hills
Distance between rows:	8 feet between hill-rows
Depth to plant:	1 inch
Seed that will germinate:	70%+
Maturity:	80–95 days
Yield/100-foot row:	30 melons

Well-drained, sandy loam with plenty of organic matter is the soil for watermelons. Mulch heavily, but make sure the ground is good and warm before you do. Watermelons grow like cucumbers. They're first cousins. They need water and plenty of hot sun. Pick when vine-ripened, and for best flavor, early in the morning when the night has cooled the melons.

GATHERING YOUR OWN SEED

Hybrid plants often do not breed true from seed. But with non-hybrid plants, instead of buying seeds every year, collect and save seeds from the best vegetables you've grown in your garden to plant. You may well develop a breed particularly suited to your region. Sometimes you'll have to buy seeds the second year, even if you decide to grow your own, simply because your plants won't produce seed till their second year. But be patient.

Always pick the best overall plants as your seeders. With crops such as spinach or leaf lettuce, where it's the leaves you want to harvest, select a plant that takes a long time to go to seed. With root crops such as radishes and carrots, use those plants that go to seed first, since this will mean earlier roots in your new plants. Here are some of the easier seeds to collect.

Beets. If your winters are mild, beets can be left in the garden. If not, store the best roots, set in sand, in a cold, damp spot. Come spring, replant the beets, and when the seed stalk develops, tie it to a stick to keep it from breaking off. Cover with cheesecloth as it matures; birds just love beet seeds. Once the stalk is mature, break off and hang it indoors with your onions. Remove the seeds when dry. With luck, you'll have seeds in time for a late crop the same year. In any case, you'll have plenty for the following year. Store in dry glass jar.

Cabbage. Cross-pollinates readily. If you want seeds to be true, grow only one species. Store the firmest, largest heads in individual baskets of soil in a root cellar or other cold spot over the winter. Handle as you do the beets.

Carrots. Same as beets, but don't try replanting if there are plenty of wild carrots in your region or you'll end up with worthless half-breeds.

Cucumbers. Let them vine-ripen to yellow-orange, even slightly mushy. Remove seeds, wash, and let them sit in shallow dishes of water to ferment for two days. Wash them again, pat dry in an old towel, and dry further by layering between sheets of newspaper. When absolutely dry, store them in a tight glass container in a dry corner of the cupboard. Check once a month or so to make sure no mold has started.

Lettuce. Same as cabbage, except the flower stalk develops in the fall of the first year so you don't have to replant the head in the spring.

Peas. Leave on the vine till the leaves turn color and begin to die. Pick the pods and shell the peas. Dry them further between newspapers in a dark spot free of any dampness. Watch for mold. Store in dry glass jars.

Potatoes. Pick your plumpest, best-shaped, scab-free ones. Store whole, buried in dry sand in your cold cellar. For planting, cut into seeds, each with an eye or two.

Radishes. Same as lettuce.

Rutabagas. Mulch heavily over the winter. Uncover early in spring and collect seeds as with beets. Rutabagas are hardy enough so that you don't have to transplant them if well mulched.

Tomatoes. Same as cucumbers.

Turnips. Same as rutabagas.

There's no mystery in gathering your own seed except for the mystery of nature. And it is truly the way to self-sufficiency and a better farm.

Pest Control

In HUMBOLDT COUNTY, California, there is a monument to a bug, a genuine crawling variety called the Chrysolina beetle. The Chrysolina thrives on St. John's-wort, a European weed that misbehaved when accidentally introduced to California. It was such a pest that it cost the ranchers tens of millions of dollars in displaced valuable grazing plants. Enter Chrysolina beetles, the good guys from Australia who like nothing better than St. John's-wort for breakfast, lunch, and dinner. In a few years they had restored the crop balance.

Not all insect activity is as dramatic in its contribution to the farm, but the vast majority of bugs are beneficial ones. And without insects there would be next to no plant life at all. So watch what you squash.

The first thing to consider in pest control is that strong, healthy plants seem to be the least tempting to most destructive insects and diseases. Perhaps these preying insects and blights are part of nature's way of improving the species, culling the weak and future problem plants. Perhaps the stronger plants are simply too tough for the pests to handle. Whatever the reasons, well-tended, organically grown crops put you halfway home in your battle with the pests.

As to the other half, there are ways and means of battling nuisances without resorting to poisonous residual chemicals. Nature strives for balance, but she works slowly. Man's disruption of that balance proceeds much more rapidly. The imbalance is caused not only by over-farming and the elimination of natural predators, but also by the accidental importation of pests from other countries to an area where they meet with no natural enemies, and by the simple expedient of crop concentration. When massive fields of one crop are sown together to increase yields, it is not only man that profits from the abundance, but the insect population as well. An acre of corn can sustain many more insects than a few wild ears. And, of course, more insects means even more the following year.

Crop concentration is necessary to feed today's world. But the answer to the pest problem it creates should be benevolent insect concentration, not poisonous pesticides. When man increases the crops under cultivation, he must help increase the insects proportionately if he is not to upset the balance of nature more than he already has.

The use of chemical pest control has been largely self-defeating over the long run. The most powerful insecticide or fungicide fails somewhere. A new species of fungus develops, for instance, that is immune to it, and a crop that has been protected by that particular fungicide for years is suddenly left wide open to destruction by the new mutant. So a new, even more poisonous fungicide is developed to deal with the new species, and eventually a new hybrid fungus escapes its clutches as well. Meanwhile, the world's fields become ever more poisoned and the chemical companies more profitable.

Wouldn't it be better to accept the fact that in some years part of a crop will be destroyed, and simply plant enough to compensate for that eventuality? The vast majority of insects, fungi, and bacteria are useful to man. Those destructive to his fields can be kept under control by growing an abundance of healthy, strong crops. In years of superabundance when no pests strike, the extra yield will not go to waste. A well-managed farm has no waste.

Always grow a little more than you think you will need. Any excess goes to your livestock and returns to the fields in the form of manure. You have better livestock at less cost, and healthier future plants from the fertile fields.

While you're willing to accept some crop loss, you will still want to minimize the effects of pests by achieving a natural balance. So you need a deterrence program too. Such a project would include introducing beneficial birds and bugs and cooperative bacteria to your farm, interplanting synergistic with insect-repelling crops, and mechanical controls like traps and safe, nontoxic sprays.

BASIC ANTI-PEST PRECAUTIONS

1. Use crop rotation, even in your vegetable garden.
2. Choose disease- and insect-resistant plant varieties wherever possible.
3. Always remove dead or diseased plants.
4. Cleanliness is vital. Keep the orchard and garden free of trimmings and waste.
5. Take care of the birds and they'll take care of you. Come to a happy agreement on the fruit crops by giving them their own to eat.

BENEFICIAL INSECTS

The two insect assistants easiest to obtain are the ladybug and the praying mantis. Both are readily available through mail order, and both "breed like flies" if conditions are right—that is, if your farm is riddled with aphids, mites, mealy bugs, grasshoppers, etc. When the pests have all been devoured, your population of ladybugs and mantises will decrease proportionately. They're a self-liquidating extermination squad.

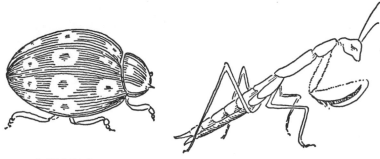

LADYBUG   PRAYING MANTIS

BENEFICIAL INSECTS

Ladybug. Soft-bodied insects are a ladybug's favorite food—spider mites, aphids, alfalfa weevils, assorted eggs and larvae. Being small themselves, ladybugs can squirm into tiny crevices, the underside of buds and leaves and the like, that chemical sprays often miss. And they don't go there to sleep! One ladybug will put away fifty or more aphids a day in snacks. Although this number might seem ineffectually small, consider the fact that most people ordering the bugs get ten to twenty thousand beetles for their garden at less cost than insecticides for the same task. Besides, one ladybug will lay up to a thousand fertile eggs.

Green Lacewing. More or less the same diet as ladybugs, with a preference for moth eggs. Also available by mail order and easily introduced.

Praying Mantis. One of the hungriest of insects. So much so, in fact, that for years it has been notorious for the female's habit of devouring the male after mating. It starts off its youth on a diet of aphids, leafhoppers, and other soft-bodied insects. As it matures, its menu expands to encompass tent caterpillars, flies, grasshoppers, locusts, and many other bugs—including bees, so watch where you put mantises in relation to your hives.

Praying mantises are poor fliers, and their distribution must be more carefully tended to than that of the other insects. Six egg cases per acre, evenly distributed, is usually sufficient. Tape the egg cases to shrubs, trees, or poles where they will be sheltered from wind, about four feet from the ground. Be careful not to cover the median band; it's their only exit from the case. Come May or June you'll have a mantis explosion on your hands.

Firefly. Fireflies are not easy to come by. But if you have them around as native residents, think of them in terms of more than a pleasant summer evening's diversion. They are particularly fond of eating your snails and slugs.

Trichogramma. When you see this one advertised, it's usually under its technical name "trichogramma" simply because it doesn't have a popular name. After all, who ever heard of a popular wasp? Well, this one should be one of the most courted of all insects, at least of those in the midget class. Its wingspread is about 1/50th of an inch.

Yet for all its lack of size, it's an extremely good destroyer of moth eggs—including codling, gypsy, luna, and royal—corn, peach, and squash borers, cankerworms, cabbage loopers, corn earworms, etc., etc., etc. One package of the usual size contains several thousand trichogramma wasps and suffices to clear five acres or so of most pests— unfortunately, it will also wipe out some beautiful butterflies.

As you can tell by the raging appetites of these insects, care is in order or you'll finish off too many beneficial bugs with your invasion corps. Use them only in areas where pests have become a problem, and don't release them indiscriminately. A natural safety check is that most of them will not migrate far. So by having a wooded area and meadows that remain off your invasion map and a sufficient distance away from the main battles, enough of the desirable insects will be spared to do their job and feed the bird and animal population as well.

BIRDS

Those barn swallows hanging around your place put down over a thousand mosquitoes, flies, and other pests a day. A flicker will polish off two thousand ants for breakfast. And a yellow warbler considers ten thousand plant lice a comfortable day's eating. Sure, the birds will eat some of your crops too, particularly the cherries. But they can be controlled. For instance, you can hang netting over the trees to keep the birds from the fruit if you want. By just planting rows of mulberries and wild cherries around the orchard you'll stop most of the raiding, since mulberries and wild fruit are their preferred dessert.

You might still lose a little of the fruit yield by attracting birds to your place—but you'll lose a lot more insects. And the birds themselves bring pleasant song and life to the farm. Build a few birdhouses and winter feeders to assure continuous occupation. Leave scraps of hair, twine, feathers, and rag near nesting areas so their life will be a little softer.

SNAKES

If you don't like snakes, remember they don't like you either. You take one road and they'll take the other, given the choice. But whatever you do, don't kill them unnecessarily. They keep not only insects in check, but also gophers, moles, rats, and other rodent pests.

BACTERIA

Bacteria often work as living insecticides. This area of pest control holds great promise of progress. For instance, *Bacillus popillae,* or milky spore disease, is one of the best controls there is for Japanese beetles, and it is being used in such diverse places as gardens and golf courses with great success. Should you suffer a Japanese beetle invasion, cut it to the quick with milky spore disease before it gets settled in.

Other bacterial benefits will no doubt be available once further research and production facilities have been established. Viruses in particular offer great hope, since they are very host-specific. The ones that kill a left-handed leafhopper won't touch even a right-handed leafhopper.

INSECTICIDE CROPS

Some insects like one plant but are allergic to others. By inter-cropping plants in alternate rows in the home garden, many pests can be minimized. The following are likely assistants in your garden and orchard.

Garlic.   Plant near lettuce and peas to deter aphids.

Geraniums or Marigolds.   Grow around your grapes, cantaloupes, corn, and cucumbers to turn away Japanese beetles and cucumber beetles.

Herbs.   Rosemary, sage, and thyme near cabbage discourage white cabbage butterfly.

Nasturtiums.   Near beans to turn away Mexican bean beetle; also good for fighting aphids in the orchard.

Potatoes.   Next to the beans to guard against Mexican bean beetle.

Tansy.   Around peach trees to deter borers.

Tomatoes.   Near asparagus to ward off asparagus beetles.

"STUFFED PESTS"

Some insects, such as potato beetles, actually would prefer something else. Say a couple of nice mouthfuls of bran. Take some dry bran and sprinkle it on top of potato plants early in the morning. Come

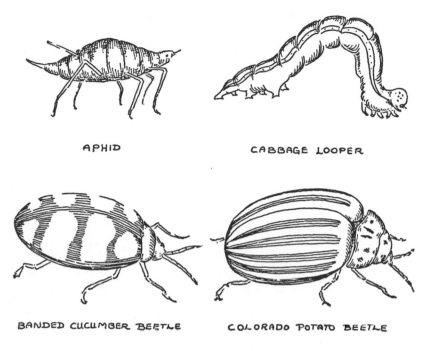

APHID

CABBAGE LOOPER

BANDED CUCUMBER BEETLE

COLORADO POTATO BEETLE

HARMFUL INSECTS

breakfast time, the bugs will stuff themselves and drink the dew to slake their thirst; stuffed and content they'll loll in the sun till their guts burst open from the water-expanded bran.

MECHANICAL DEVICES

Make insects walk the plank. Take a small board and coat the bottom of it with molasses. Lay it on the ground near the plum trees just before and during fruiting. In the morning turn it over and destroy the plum curculio beetles that have decided to take shelter there. The curculio beetle is about a quarter of an inch long, brown overlaid with mottled gray, and has four small humps on its back. Its larvae are those worms you find in cherries, plums, and other stone fruit.

If you want to make a larger operation of it, spread a canvas beneath the tree. Pound the tree trunk with a rubber mallet. If you've got curculios, they'll come raining down. They'll also play possum long enough for you to dispose of the lot.

Banding traps are simply bands of sticky material wrapped around tree trunks—sort of waist-high, horizontal flypaper. A common one is tanglefoot.

Codling moths and apple orchards are to some extent inseparable. To minimize the problem, remove all early-fallen apples, that is, the ones that aren't really ripe, and burn them. These apples contain larvae of the codling moth. By removing and burning the apples before the worm pupates, or splits and digs in for the winter, you break the breeding cycle and for the most part eliminate this pest in the coming season.

Lights will draw large numbers of insects at night. But light-trapping is not recommended except in real emergencies, because this form of trap attracts almost anything that flies, and the ratio of beneficial to harmful insects killed is too high.

BEER AND NONTOXIC SPRAYS

If you're bothered by slugs in your garden, consider beer. Pour some out in a dish and place it between rows of plants. Slugs, as indicated by their crawling form of locomotion, are lushes at heart. They will come from all over for your beer. However, they also can't hold their alcohol. Simply fish out dead-drunk, drowned slugs every morning.

Mention sprays, and the first thought that comes to mind is the ubiquitous chemical industry. But how does water sound as a safe spray? Spray off aphids, mealy worms, and cabbage worms with water pressure from an orchard sprayer. Half of them won't go back to the plant. Well then, you ask, why don't commercial farms use water instead of insecticides? The answer is simple. Water has no residual effect. You'll have to spray again and again if you're saddled with pests.

Use a water spray on the loose bark of your fruit trees as well, to wash out whatever's there. Hold the sprayer two feet away and concentrate on the trunk, large branches, and crotches. This helps keep down codling moth.

Miscible oil sprays are one of the best protectors your orchard can have. These should be used only during the winter and very early spring while the trees are dormant, however. The spray seals a tree in a microscopic envelope of oil. The tree, being dormant, doesn't need more oxygen than it can take in through its roots. Not so all those insects hiding beneath the bark. They're smothered. Apply with a regular orchard sprayer. Cover the entire tree in one thorough spraying. Dormant miscible oil sprays can be purchased, or you can mix your own.

DORMANT MISCIBLE OIL SPRAY

Mix:
- 1 gallon light-grade oil
- 1½ lbs. fish oil (or 2 lbs. regular) soap
- 1 gallon water.

Bring to a boil. Pour back and forth between two pots. The soap will act as an emulsifier and let the oil mix with the water.

Dilute with:
- 20 gallons water.

Use as soon as possible. If the mixture separates between uses, reheat and mix again.

Of all the treatments for pest problems, however, the most effective is to minimize them in the first place. Prevention is the best cure.

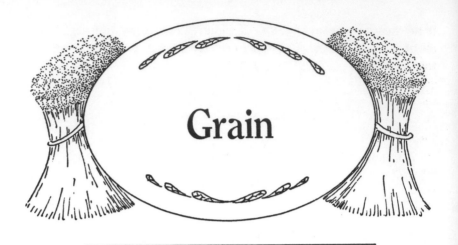

Grain

*A police state finds it cannot
command the grain to grow.*
—JOHN F. KENNEDY

T HE ROLLING FIELDS of golden wheat that you think of when you hear the word "grain" are basic to any agricultural venture. But they come in small, easy-to-care-for homestead sizes as well as in the huge horizonless acreage of the plains. And you should have a few such fields if your farm is to be in any way self-sufficient. For it is these field crops, cultivated for their seeds, that supply our bread, and much of the food for our livestock. Without the cereals and related field crops, civilization—perhaps even language—might never have developed. It is the field crops that persuaded man to settle down and take up the plow.

Today's cereals are products of domestication as much as the cow or the dog. They bear little resemblance to their ancestors. Wild corn, whose seed served only the natural function of reproducing the species, may have yielded no more than a quarter of a bushel of small, hard kernels per acre. Today's cultivated corn has been known to yield over 150 bushels of large plump seeds per acre.

The beauty of it is that for the most part this domestication has been a natural process. Somewhere, sometime, someone found an ear of wild corn that was a little bigger, a little juicier or sweeter than the rest. Instead of eating it right away, he thought maybe if he planted it the new corn would be·a little better too. It probably wasn't—not all of it, at least. But a few plants, maybe even a quarter or half of them, were. He saved the best of these for seed again—and so on for generations. In recent times hybridization, still a natural process of selection but guided directly by man, speeded up the results. The development of modern cereals, for all its lack of publicity, far outshadows that of the hydrogen bomb. Without these grains the world could not possibly be fed any longer, whereas it could still be destroyed quite effectively without the aid of the H-bomb. Of course, you're not going to save the world with your small field of grain. But choose your seed carefully, as the farmer with a reverence for nature has done since he first tilled the soil, sow, cultivate, harvest it well— and your bread will be wholesome and good, the feed for your animals nutritious, your farm a small but indisputable proof of better things to come.

Fine, you say, but where do I start? Well, you'll want wheat, and maybe buckwheat and rye, for your bread, oats and corn for your animals, and no doubt a few ears of sweet corn for yourself. The amount of land needed for this is surprisingly small. Just a half-acre of wheat and one acre of corn will take care not only of your bread box, but a small, healthy flock of chickens as well. An additional three or four acres of corn, oats, and a hay mix that grows well in your region will round out your grainery enough to feed some pigs and goats. A two-acre pasture will take care of them through the summer. If you happen to be a brown-rice addict, incidentally, consider cultivating a taste for the other grains—a rice paddy in this country is not something for the apprentice farmer to tackle.

BUCKWHEAT

The amount of buckwheat grown in the United States is relatively minor, perhaps 1/2000th that of corn. Still, it's an important crop for the small farm, not only because buckwheat flour makes the kind of country-morning flapjacks you dream of, but also because if you keep bees, the rich dark buckwheat honey you'll get is great on top of those pancakes. The middlings, which is what remains after milling flour, make fine stock feed, and the whole grains are excellent for the chickens. On a commercial scale buckwheat is processed for rutin, a relatively new wonder drug that combats hemorrhage, frostbite, gangrene, high blood pressure, and even to some extent radiation damage.

Buckwheat is a fairly new crop thought to have originated under cultivation in China about a thousand years ago. It is an erect herbaceous annual usually reaching a height of three feet or so. The alternate leaves are triangular, varying in length from two to four inches. The single taproot, though it has few branching roots, is nevertheless a very effective extractor of soil nutrients, and the buckwheat plant can avail itself of minerals sometimes unobtainable by wheat, oats, and other grains. It also improves soil conditions more than these. Hence the old saying of some farmers that if the land is too poor to grow anything on, it will still carry buckwheat. It's an ideal crop for a field in need of revitalization.

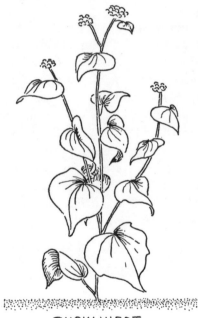

BUCKWHEAT
GROWS 1-2' HIGH

ADAPTATION

Buckwheat is best grown in the northeastern part of the United States, although its range could probably be extended considerably. If it's not grown in your region, experiment with a small lot, as long as the physical conditions seem appropriate. It does well on most well-drained soil, better in a cool, moist climate, but is susceptible to cold. Since high temperatures and dryness cause the plant to set seeds very poorly, it is usually sown as a fall crop. Liming is not necessary, the plant preferring slightly acid soils.

TYPES

The two main varieties of buckwheat grown for flour are the Japanese and the silverhull, which are often sown mixed. A third variety called common gray is also used with some frequency. Other varieties you may run across are Tartary and Tetraploid; these, how-ever, are grown mostly for rutin production and not for their edible end products.

SEEDING

Drilling is preferred. Use three to four pecks of seed per acre in seven-inch rows, planting twelve weeks before the first expected killing fall frost. This assures that the seeds will set in cool weather. Broad-casting the seed at four to five pecks per acre will probably give you a good enough crop if you don't have a grain drill. You can expect to harvest between twenty and thirty bushels an acre.

HARVESTING

Harvest in late fall just after the first-formed seeds mature. One of the problems with buckwheat is that the seeds on the plant mature at different times; therefore harvesting is timed to maximize the num-ber of mature seeds. The plants are cut, shocked, and field-dried. Field drying takes about ten days. Then the buckwheat tops are ready to be stored whole for winter fodder—chickens will pick out the seed as you feed it to them—or threshed like wheat if you're planning to mill your own flour.

CORN

It's corn that is as American as apple pie, not apple pie. Apples are a European crop. Corn not only originated in the Americas, but most early American civilizations were based on it. The Aztec, the

Inca, and the Mayan were all corn cultures in the same way that the Khmer is a rice culture and the European a wheat culture. Even today it is a keystone crop of Western hemisphere agriculture. But it's not the same sweet corn you grow in the vegetable garden. Field corn is grown for its storage quality; if you try to boil it up for dinner, you'll be very disappointed. Still, if its woody flavor doesn't appeal to you, the livestock love it.

Corn is a highly developed annual grass. It has been so domesticated into a superproducer, in fact, that it can no longer survive in the wild state. Wild corn is extinct.

The corn plant is composed of a tall tapering stem as thick as one and a half inches at the base and as tall as eighteen feet in the case of some hybrids grown under ideal conditions. Rooting is strong, particularly the development of brace roots when the plant reaches its maximum growth. However, corn roots do not improve the soil. The crop is very hard on soil in general. Never plant a field to corn unless it has had two years of another crop in rotation.

ADAPTATION

Although corn is of tropical origin, it does well in subtropical and temperate climates. It's the summer temperatures that count. Since it is an annual, even subzero winters will have no effect simply because the plants aren't around during that time of year. They're seeds in your grain bin.

Enough moisture is the big problem with corn. Almost any soil will do, but if the rainfall is insufficient, the plant will not thrive. This is particularly important during flower fertilization. Your corn crop will do best if there is a fair amount of rain the four weeks before silking—the stage of development where pollen appears on the tassels topping the stalks and the silks appear on the incipient ears of corn. Rain a few weeks later will give it an extra boost.

TYPES

Corn comes in two basic colors: white and yellow. Since yellow corn contains much more vitamin A, it's preferred. Multicolored corn, known as "Indian corn," is usually grown only for decoration, although you will sometimes find freckled ears in other varieties. Corn is usually classified by its type of kernel and starch.

Dent Corn. The most widely cultivated. Each kernel is composed of both hard and soft starches. The soft is in the center and toward the top. It shrinks upon drying and the distinctive dent forms at the top of the kernel. For a change, try dent corn instead of popcorn some winter evening. It won't explode, but will rupture more gently, the kernel about doubling in size. Heated up in oil like popcorn, it's known as "parched corn," and tastes like those delicious, crunchy, half-popped pieces you used to get toward the bottom of the popcorn bag at old grade-B movies.

Flint Corn ("Indian Corn"). Grown in cooler climates and high altitudes because of its early-maturing quality. Kernels contain soft starch in the center with a coating of flint-hard starch all around. Variegated kernel coloration.

Flour Corn ("Squaw Corn"). The all-soft starch of the kernels lends itself to flour milling, just like wheat, although it will be a rougher flour. It is also known appropriately enough as "soft corn." White and/or blue kernels.

Pod Corn. A primitive corn in which each individual kernel is enclosed in a pod, or husk. A very leafy plant often used as forage corn.

Popcorn. What makes this corn the stormy evening fireside companion it's famous as is its hard endosperm, or outer covering. When the corn, with a moisture content of roughly 12 percent, is heated, the endosperm prevents the water vapor from escaping. As pressure builds up, this outer cover ruptures with a pop.

Sweet Corn. What you raise in your vegetable garden, not for the animals—unless you feel extravagant. Kernels are translucent and smooth, and contain sugar as well as starch. Much of the sugar is broken down within a few hours of picking, which is why from-the-stalk-to-the-boiling-pot sweet corn makes the supermarket variety taste like rejects from a birdfood manufacturer.

Waxy Corn. The endosperm is composed of a starch with a waxy appearance, particularly when cut. It is an industrial corn in the sense that the cornstarch derived from it is used primarily for making adhesives and sizing for textiles and papers.

Hybridization has expanded the choice even further, and is the main thrust behind the phenomenal corn yield increases in the last decades. However, if you're thinking of growing a hybrid corn, note that you cannot save the seeds from a double hybrid, that is, one produced by crossing four corns instead of the two used in common hybridization, for future planting. The seeds from double hybrids will not grow true. There's something about double-crossing that makes them revert to type. Most of the corn planted today is hybrid, and even single-crossed varieties grow best from primary seed stock, or that produced by the actual hybridization. So if you want to plant your next year's corn from this year's seed, you'll have to do some shopping around for nonhybrid seed.

SEEDING

Corn is usually planted one or two weeks after the average date of the last killing frost of spring. In southern Texas this means as early as the end of January. The Corn Belt usually plants sometime in May, depending on the area. Location of your farm and its land contours will be the deciding factors for planting time. Some farmers will plant sweet corn ten days earlier than everybody else in the vicinity because they have a southern hill much less exposed to frost than the surrounding area. For each two days of earlier planting they get one-day-earlier mature corn. This means they can take their sweet corn to market as much as a week before the rest of the farmers. The price they get for it is much higher. So, of course, is the risk they take. One late frost could wipe out the crop. Still, there's more than just the grocery money to "first cornitis"—there's that certain irresistible pride in having the earliest sweet corn of the season.

Germination of corn seed is best at a soil temperature of 60°F. At 55°F. it is retarded, at 50°F. or below it is minimal. The more clayey the soil, the more important the temperature. If your soil is loamy or sandy, germination will be good at temperatures as low as 55°F. Soil temperatures can be measured directly with a thermometer. But unless you're experimentally inclined, just plant when your neighbors do; they know the timing instinctively.

Planting corn about an inch or two deep ensures it enough moisture. Planting it deeper has not shown any increase in yield. Generally, the heavier the soil, the closer to the top it should be seeded.

Open soils need planting at the full two inches, since they give up moisture more readily. Plenty of organic matter in the soil is, of course, a plus.

Corn can be drilled, check-rowed, or hill-dropped. Drilled corn usually outyields the others and is the easiest, so stick with this method. Seeding varies from seven thousand to twenty-four thousand plants per acre. In most cases there is little yield increase with over sixteen thousand. Another big variable is the width between rows— usually forty, thirty, or twenty inches. Rows thirty to thirty-six inches apart give the best results, provided your soil is in good condition. If it's not, switch to wider rows.

Chances are you'll have someone else come by with a corn planter till you can get your own. In that case, don't be too insistent on having your own way. If he spaces his rows at forty inches, don't demand twenty inches or you might not get your corn planted. There will be a difference in yield, but only of 5 percent or so, a major factor on the large-scale farm but not on the small one. To boot, in some areas there are positive benefits to be derived from spacing as wide as sixty inches, so listen to the old-timers' advice.

Seven to ten days after planting, the seedlings will break ground. While you might not hear the corn grow—then again you might—its progress is very rapid. Mark the top of one plant's growth on a stick early some day after a night of rain. Come back after supper for a surprise.

HARVESTING

You'll probably find the smallest farms in your area utilizing a corn picker, even if it means renting one or bringing in a custom man. It is a significant time-saver. Picking an acre's worth of corn takes an experienced man about eight hours. A two-row picker will harvest and husk it in under two.

With only an acre or two of corn, you'll have little difficulty hand-harvesting, even with the seventy-five to one hundred bushels of corn on the cob you can expect to pick per acre. But remember, when you get to the husking stage, your hands are going to feel it unless they've built up some good hard calluses from other farm work.

As opposed to your small plot of sweet corn, which is harvested by the ear the very day it ripens sweet and milky, feed corn is harvested when the upper leaves have lost half their color and the lower ones all of it. The husks should be dry, and in the case of dent corn, the dent, appropriately enough, should be well dented. Harvesting is usually best after a frost.

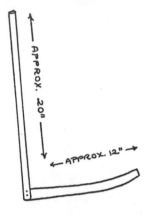

APPROX. 20"

← APPROX. 12" →

USED TO CUT THE WHOLE PLANT AT THE BASE BEFORE SHUCKING THE CORN. SWING IT MORE OR LESS AS IF YOU WERE PUTTING WITH A GOLF CLUB.

CORN KNIFE

If you intend to use the cornstalks as bedding for your livestock, cut individual stalks with a corn knife. Bind ten to fifteen stalks into a bundle and shock ten to fifteen bundles together. Let the shocks field-dry for a couple of weeks. The precise drying time will depend on the weather. Ears will break off crisply from their stalks when ready. Haul the shocks of corn to the barn and stack them. Pick the ears off when you have time.

If you plan on plowing under the following spring (remember to go over the stalks with a disc harrow before plowing), drive a wagon into the field when the ears are ready to be picked and do just that. Haul the dry ears back to the barn.

Husk the corn for animal feed the way you would sweet corn for the pot. It will be easier, because it's dry, but count on getting those calluses all the same. In the old traditional husking bees whoever drew a red ear of corn got to kiss the partner of his choice. So the clever farmer slipped some Indian corn into his planting to assure eager workers. Ten percent red ears is about maximum, however, if you intend to get any work out of your crew, and, as one farmer put it, it takes a lot of advance planning to make sure the right people show up.

Pigs will eat whole corn on the cob happily. For your chickens, however, it must be shelled. This is done by running the cob through a sheller—by hand it would take forever. A hand-crank model suffices for the small farmer. For goats and other ruminating livestock, the best use is made of corn when it is ground whole, cobs and all. Take it to your local mill for grinding.

OATS

The world's most important crops are wheat, rice, corn, and, coming in fourth, oats. As animal feed, oats were once primarily the domain of horses. However, with the decline in the equestrian population, they have been used more and more as a general livestock feed, for dairy cattle, young sheep, hogs, and other animals. Oat hulls, the by-product, are used commercially to make furfural, a solvent and chemical intermediate widely utilized in oil refining, pharmaceuticals, and the production of nylon. For the small farm, if growing conditions permit, oats are highly recommended, not only for their food value, but also because the straw makes the best bedding of all grasses. And they are a good nurse crop for a forage field of alfalfa.

A nurse crop is one that grows quite quickly and is sown together with a much slower-growing one to provide shade and minimize weeds. When oats are sown as a nurse crop with alfalfa, the oat crop is harvested as it matures in summer. Then in fall the first alfalfa crop can usually be harvested from the same field. The alfalfa is harvested again the next year, maybe three or four times in a season—just like mowing a king-sized lawn.

ADAPTATION

Oats are a moisture-demanding cereal. They are particularly prone to poor yield if hot, dry weather predominates during the latter part of their growth and kernel development. Oats do poorly in high altitudes (over two thousand feet) and best in cool, moist climates. About soil they're a lot less fussy, however. Almost any well-drained soil will do.

TYPES

The five general types of oats, classified by the color of their hulls, are black, gray, red, white, and yellow. The most common in the North is the white; that preferred in the South is the gray or the red. You'll want to start with a regionally tested variety.

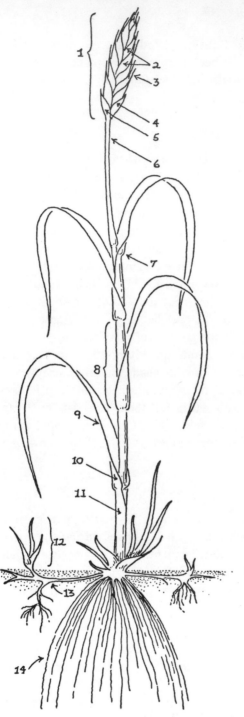

PARTS OF GRAIN-BEARING
GRASS PLANT

1. SPIKELET
2. FLORETS
3. AWN
4. SECOND GLUME
5. FIRST GLUME
6. CULM
7. NODE
8. INTERNODE
9. BLADE
10. COLLAR
11. SHEATH
12. SHOOT
13. RHIZOME
14. FIBROUS ROOTS

SEEDING

The more your oat seeds weigh, the better their quality. A heavy bag of oats means light, poor-yielding seeds have been culled out.

The amount of seed per acre you can sow will depend on the rainfall figures for your area. With a lot of rain, there will be less moisture competition among the plants, so you can have more of them to a field. In drier areas of the country, such as the Great Plains, four to eight pecks per acre is common. Land in the eastern states can usually carry a seeding of six to ten pecks. In the rain-laden Pacific Northwest the amount can be raised to ten to fifteen pecks. The heavier end of the seeding scale is used in fall planting to insure that a good crop is sustained through the winter for spring harvesting.

Oats may be fall- or spring-sown. Spring sowing predominates except in the more southern areas, where winter oats, along with all their other attributes, make an excellent cover crop. Fall seeding should be done early, usually thirty days before the first expected killing frost; spring seeding one week before the average last killing frost.

As is true of most grain crops, drilling will give bigger yields with less seed than broadcasting. Plant your oats in drills four inches apart and cover with two inches of soil. Broadcasting might yield only twenty-five to thirty bushels of oats an acre as opposed to the forty or fifty obtainable from the same field drilled. Then again, you're not really out to set world records. If you can't get someone to give you a hand with his grain drill in exchange for part of your crop, just increase the size of your planting a little.

HARVESTING

A crop of oats is harvested when the kernels are in the dough stage—not soft enough to squash between your fingers like a bug, yet soft enough to be dented by your thumbnail. The heads will be yellow, about half the leaves still green. If a combine-thresher is used, harvesting is done about a week later, when the oats are fully ripe. But with hand-harvested oats, it's better to harvest a little early than a little late. Cut with a scythe and grain cradle. Windrow it, that is, rake it into long, narrow rows, up and down the field, with about eight to ten feet between the rows so the wind can dry it out. Let the windrows lie for a day, then bind into bundles eight to twelve inches in diameter.

Lean several bundles together so they stand upright. The oats should field-cure for about two weeks, then be threshed like wheat. Feed the whole oats to the livestock, or grind for oatmeal.

RYE

Being a rye bread nut myself, I can't recommend too highly growing this one. Besides, rye is easier to grow than wheat, if not as high-yielding, and makes excellent green manure. Rye is one of the more recent cereals to come under cultivation. It is hypothesized that it grew only in the wild state as late as twenty-two hundred years ago. Even so, its new-found popularity is already declining. Today it is grown more for green manure than for flour; barely a fourth of it is harvested for grain. People seem to prefer the white chemical cotton that passes for bread these days to a solid, dark, peasant loaf that is naturally wholesome. Another possible reason for the decline of rye as anything but a green manure crop may be that the main use for rye straw used to be in making horse collars.

To give you an idea of what you can expect if you decide to grow rye for flour, rye weighs fifty to sixty pounds a bushel and you can expect to harvest twelve to eighteen bushels an acre.

ADAPTATION

Rye will grow in almost any soil, including sandy, acid stuff that will not carry most other grain plants. It is this ability to thrive on relatively infertile soils that makes it such a valuable green manure crop. It requires less moisture than wheat and much less than oats, but will also grow well with more. Rye can be planted in any state in the United States. Over 50 percent of the commercial rye crop in the United States comes from the northern Great Plains states, where the limited rainfall makes it an ideal crop.

TYPES

Dakold. For the coldest regions, such as Montana and the Dakotas.

Rosen. Middle-temperature regions.

Abruzzi. For the southern parts of the country.

SEEDING

Pick a variety suitable to your region. Growing Abruzzi up north, for instance, can lead to almost total crop failure—except, of course, rye is never a total flop, since there will always be something there to plow under as green manure.

Rye, like wheat, is sown about the time of the average first and last killing frosts of fall and spring, respectively. The exception to this is winter rye sown for pasture or for green manure, in which cases it should be planted a month earlier. Planting for green manure requires more seed than planting for grain, since you're after the succulent matter rather than the flour. For manure, use up to two bushels per acre; for grain, three to six pecks, depending on the moisture in your region. The wetter the weather is apt to be, the more seed you can plant on a field. In the same vein, the moister the ground, the shallower the seeding, the usual range being half an inch to two inches. Seeding deeper than you need to reduces the yield.

HARVESTING

Rye, like oats and wheat, is also harvested when the seed is in the dough stage. Treat the same as you do wheat, including milling. But remember, if you have a field of wheat and one of rye, the rye will ripen up to a week earlier.

SORGHUMS

One doesn't hear much about sorghums in the city, probably because you can't serve sorghum on the cob. Essentially, this crop is the more arid farm's corn. It's not that sorghums need any less water than corn, it's just that they can extract it from the soil more efficiently and are more drought-resistant. If you live in a dry part of the country, many around you will no doubt be growing sorghums. But avoid the crop if you can. Sorghums contain prussic acid, and while careful preparation will make good silage of it, pasturing on it, unless you're absolutely certain no young plants are present, will often kill the livestock. Poorly prepared silage will do the same. Sorghum hay causes no problems, since the prussic acid is removed in curing. But along with their other drawbacks, sorghums are hard on the land, so why get involved?

SUNFLOWERS

Usually considered a forage crop, the sunflower is not a grain, and is included here because I happen to like eating sunflower seeds almost as much as corn on the cob. So do chickens, and it makes as good a poultry feed as any of the grains.

In some parts of the country sunflowers are used for silage. However, they give it a resinous taste not favored by livestock. For this reason it is disguised for them as molasses silage.

Grow some sunflowers for yourself and the chickens. A field of eight-foot-high sunflowers, with the flowers themselves up to a foot in diameter, is a fabulous sight.

ADAPTATION

Sunflowers grow almost anywhere in the United States. They are particularly useful for "grain" production in high-altitude regions where low growing-season temperatures tend to give poor corn or sorghum yields. Sunflowers are also more resistant to frost, making them useful in areas of short growing seasons. They thrive on good soil, but can also be grown on poorer ones. Highly variable in yield, with a range of anywhere from ten to eighty bushels per acre.

TYPES

Almost all sunflowers grown in the United States are of the Mammoth Russian variety.

SEEDING

Plant sunflowers one week before the average last frost. Use rows thirty-six inches apart if you're drilling, with six to seven pounds per acre at a depth of one to two inches. A corn planter or a grain drill with the holes blocked to give enough spacing between the rows can be used. For small lots, broadcasting will suffice, if you can get an even spread. Don't try to increase the seed quantity much above eight pounds per acre. Give them room, lots of room. Sunflowers are particularly sensitive plants when it comes to crowding and competition. Cultivation is important for the same reason; weeds will do your "turn-to-the-suns" no good.

HARVESTING

Sunflowers are harvested rather like corn. Wait till the yellow rays have about half fallen off the flower head. Lob off the tops with about a foot of stem and haul them in from the field in your wagon. Hang them by their stems, tied in bundles. When the flower heads are dry, rub them back and forth over some one-inch mesh screening fastened atop a barrel to collect the seeds.

WHEAT

The cultivation of wheat is older than history. Even in the Stone Age it was being grown by the lake dwellers of Switzerland. Nothing about its origin is certain except that it was not indigenous to the Western hemisphere.

For the bread-maker, a bushel of good wheat weighs a minimum of sixty pounds. You should get at least fifteen to twenty bushels per acre —that's a lot of bread. But you don't have to eat it all; the grain is excellent for livestock and the straw makes good bedding as well.

ADAPTATION

Wheat grows in the wide band from sixty degrees north latitude to forty degrees south latitude, doing best with a cool, moist growing season followed by a warm, dry period for ripening. There are two groups: winter and spring. Spring wheat is grown in the more northern regions of the United States, winter wheat, sown in fall, not winter, in the South.

TYPES

Hard Red Winter. A hard, high-protein wheat grown in the Great Plains area. Good for making bread flour.

Soft Red Winter. Grown in areas of higher rainfall, milder winters. Produces a soft, lower-protein flour usually reserved for sweet pastries, cookies, cakes, etc. Predominantly grown in the Midwest— Ohio, Illinois, and Indiana.

Hard Red Spring. The highest-protein wheat for bread flour. Grown mostly in the north-central states where the cold winters preclude the growing of winter wheat. Can do with less rainfall than the winter wheats.

White. Another pastry-flour wheat, this is grown in the far western states—California, Idaho, and Oregon.

Durum. The wheat with the hardest kernels. As such, its primary use is for pastas, such as macaroni and spaghetti. A spring wheat grown chiefly in the north-central area.

The growing areas and uses of a particular wheat are not rigidly fixed. But if you can grow, say, durum and hard red spring in your region, you'd be better off with hard red spring for your bread flour. Within the main classifications of wheat, just given, there are almost 250 varieties presently being grown. Local availability and planting conditions should guide you in your choice.

SEEDING

With all seeds, quality pays. Wheat is self-fertilizing, and once you raise a crop, you'll have your own seeds for the next year's planting. You'll have to cull the batch you plan to sow, picking out any weed seeds and trash that might have been collated in with it during harvesting, but the wheat seeds themselves will be of the quality of their parents.

Your first year, start with a variety of wheat that has already shown its adaptability to your region. Seed at the rate of four to six pecks an acre. If you sow late, add an extra peck or so. Drilling makes for better germination than broadcasting and is used almost exclusively on large farms. However, until you get a grain drill, broadcasting will probably give you fine results. Seed the deepest on sandy and other loose soils, the shallowest on clay soil. The range of soil cover is one to three inches, the average two.

Winter wheat should be sown about the time of the average first frost. Seeding too early increases the likelihood of lodging and of destruction of the crop by Hessian fly. But the wheat should be given enough of a chance to get started before winter sets in—underground, that is, you don't want it to start sending up surface shoots, or it will be winter-killed.

Whereas winter wheat is generally seeded on the late side, spring wheat should be sown as early as possible. Again, the date is gauged by frost time; plant at the time of the average last killing frost. Late planting by only two weeks can cut the yield of a wheat crop by a fourth.

HARVESTING

Wheat, like oats and rye, is harvested when it's in the dough stage. This doesn't mean instant bread. You still have to cut, bind, shock, thresh, winnow, and grind it. Harvest when the straw is just beginning to turn yellow and a kernel is soft enough to dent with your thumbnail, not soft enough to squash like a bug.

First the wheat is cut. You can do a passable job with a sickle on a small field an acre or less in size. A scythe with a grain cradle saves a lot of work and will handle up to ten acres of any small grain such as rye, oats, or wheat, if you're in shape. Anything over ten and you'd better look around for a custom man with a combine.

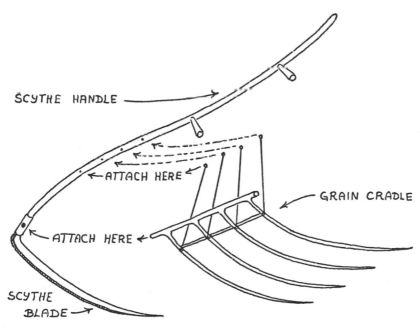

SCYTHE WITH GRAIN CRADLE

After cutting, the wheat is bound in bundles about eight to ten inches in diameter, and the bundles clustered in shocks to cure. Field-curing wheat takes about ten days.

There's a lot to be said for the old hand-harvesting and threshing tradition in terms of really feeling like you're living with the land, but if what looked at first like a manageable crop to harvest by yourself seems to be taking you forever, remember it takes time to learn new skills. Remember that too when you first size up the field. A combine run by a neighbor or custom man, cutting and threshing the wheat in a once-over operation, reduces harvesting time to a tenth of what it would be otherwise. And that saved time may be more needed elsewhere on your farm. Balance things out before you tackle the big jobs single-handed.

DAYS FROM SEEDING TO HARVEST OF GRAINS

Wheat:	90–110	(Fall sown 210–240)
Rye:	85–110	(Fall sown 200–230)
Oats:	140–150	(Fall sown 220–260)
Corn:	120–140	
Buckwheat:	90–110	

THRESHING AND WINNOWING

When your wheat is cured, haul it in off the fields for threshing. Spread out a layer of it six to eight inches deep on a clean, dry, hard surface, preferably concrete. Now take out all your frustrations on the wheat by beating it with a flail, a long pole jointed to a shorter pole which hits the grain to be threshed.

Threshing is going to be a bit of work. You're done when the grain heads are empty. Remove the straw by lifting it off with a pitchfork, using a sifting motion to let the grain drop through. Shovel up the grain and chaff and winnow it by pouring from one basket to another—try to arrange to have a cooperative wind blowing. Let the grain fall from at least four feet to permit the chaff to be blown clear. A more efficient means is a fanning mill. However, in most cases, it will be too expensive for the amount you'll be harvesting. Give the chaff to your chickens as bedding, and they will pick out any grain you missed. The wheat is now ready to mill.

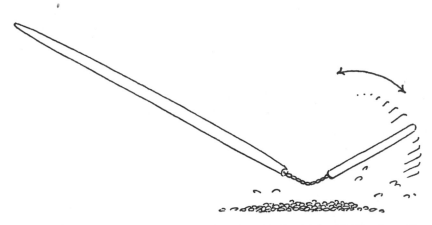

THRESHING FLAIL MADE FROM OLD SHOVEL HANDLE AND
DOWEL ATTACHED WITH CHAIN TO MAKE FLEXIBLE JOINT

FLOUR MILLING

For whole-wheat flour in small ready-for-the-oven-bread batches,
you can use a manually operated grist mill and grind your own just
before baking. For larger quantities, ferret out the nearest miller from
his floury bin and strike a bargain with him. There'll be a lot of other
equipment your farm will need more urgently than a full-sized flour
mill capable of grinding hundreds of pounds of flour at a time. Some
sunny day, for a touch of real country life as the early farmer knew it,
you might want to go hunting for a nice flat, smooth rock from the
bottom of the creek and try stone-grinding your flour on it. But don't
plan on making a general practice of bagging your own, or you'll never
get the goats milked and the eggs gathered and the honey into jars.

Forage

I believe a leaf of grass is no less than
the journey-work of the stars. . . .
—WALT WHITMAN

Forage crops are those plants raised for your livestock to eat whole. They can be stored as hay or silage. As a pasture crop, the animals eat forage "on the hoof," so to speak. It's a no-harvest, no-muss, no-fuss way of feeding them. At the same time, good pasture builds good soil, again with less work than many other methods. Just to give you an idea of how important pasturage is to animal husbandry today, over half the total feed for livestock comes from this source.

Pastures are used for grazing, and meadows for harvesting hay crops. The most economical way, in terms of labor, to feed your animals is to let them forage for it, and most of the year grazing on pasture is your best bet. However, there's the winter to contend with, and, in the more arid parts of the country, dry spells when the livestock can't get enough nourishment from free grazing. Enter hay, crops cut and dry-cured for storage—or, in more and more cases, silage.

PASTURE AND MEADOW

A pasture is a simple enough thing: an open field covered with plants suitable for animal fodder. Livestock can be grazed in wooded areas and brush country too, where conditions permit, but good wood-land clearings are hard to find, and can usually be put to better use. A native pasture is one which to all intents and purposes remains wild, or uncultivated, where you'll find a mixed bag of plants native to the region. An improved pasture, on the other hand, has a planned layout of specific crops on tilled, or cultivated, land.

PASTURE MANAGEMENT

Permanent Pasture. As the name implies, permanent pasture is land set aside for long-range use. Seeded to perennial crops or self-seeding annuals such as bluegrass, white clover, redtop, and lespedeza, the land is maintained five or even ten years solely for grazing. Acreage intended for use as permanent pasture should be sown with mixed seed. A one-crop diet is boring, not to mention being less nutritious than a full-course meal.

Rotation Pasture. Land-use rotation is practiced on all well-managed farms. When acreage is taken out of cash crop production for a rest, it is sown with pasture crops. For instance, say you've grown corn on a field one year. After harvesting the corn, you sow it for pasture and keep it that way two or three years before plowing under and raising corn on it again. Plants for rotation pasture often include alfalfa, ladino clover, timothy, and trefoil. The mix most valuable for a given area will vary with region and soil conditions. But whatever the case, a combination of legumes and grasses is almost always better than either alone.

Check with your neighbors to learn how they manage their pasture and hay meadow. Needless to say, some farmers will have better-quality livestock than others. They're the ones to discuss your pasture management with.

You'll quite often find that a neighboring farmer is willing to rent your pasture for his own use, either as grazing land or for hay. Since you won't start off your life in the country by doing everything at once, this kind of arrangement will often work to your benefit. A well-main-tained grazing pasture or hay meadow will build up your soil for future use. Besides, while you won't get rich on the bargain, you'll be getting some pocket money, or some old farm equipment, or part of the crop in return.

TYPICAL FIELD AND FORAGE CROPS FOR YOUR REGION

1	2	3	4
ALFALFA	ALFALFA	ALFALFA	ALFALFA (UNDER
ALSIKE CLOVER	ALSIKE CLOVER	ALSIKE CLOVER	IRRIGATION)
CORN	CORN	BUCKWHEAT	BERMUDA GRASS
LADINO CLOVER	OATS	CORN	SUNFLOWERS (HIGH
OATS	RED CLOVER	LADINO CLOVER	ALTITUDES)
RED CLOVER	RYE	OATS	WHEAT
REDTOP	SUNFLOWERS	RED CLOVER	WHITE CLOVER
RYE	SWEET CLOVER	REDTOP	
SUNFLOWERS	WHEAT	RYE	
SWEET CLOVER		SUNFLOWERS	
TIMOTHY		SWEET CLOVER	
TREFOIL		TIMOTHY	
WHEAT		TREFOIL	
		WHEAT	

5

ALSIKE CLOVER
BERMUDA GRASS
CORN (MARGINAL)
KUDZU
LESPEDEZA
OATS
RYE (COVER CROP)

Bartering is more fun than dealing with monopoly money, and it quite often gives the small farmer a better deal. Good land doesn't become outmoded by next year's version. And perfectly good, usable equipment is often replaced with newer, more complex equipment simply for the sake of efficiency, which means that while the older tractor, small manure spreader, or what have you is no longer practical on a large, streamlined commercial farm, its size and work load are probably just fine for yours. Everyone gains by bartering—just remember your trading partner may have more experience at it than you.

FERTILIZATION

Pastures and meadows are usually treated about the same. Both should be limed. Since hay crops are often grown as rotation crops, some farmers take a short cut and fertilize the field only when it's to be used for a cash crop. Their reasoning is that no crop uses up all the fertilizer that is applied, and so the corn or other cash crop receives the benefit of fresh fertilization, while the hay, which will be used only to feed livestock, can make do with what's left. This is true to some extent, but the hay will not do its best under these conditions. You're better off fertilizing the meadow as well as the pasture.

After all, hay is not something you grow on a spare field, unattended. Not if you expect to get the kind of quality that will make your goats, for instance, top milk producers. It's going to be their winter fodder. Like any other crop, hay needs fertile soil. This is something to remember if you plan to buy your hay from a local farmer rather than grow your own. You'll have to know the difference between good and bad hay, and the place to start is with the fields themselves. The amount of manure, potash, phosphate, and lime needed to grow a good hay crop varies. The best way to get the specifics for your particular area is by talking with your county agent and the boys down at the feed store. Whether you grow your own or buy your hay, you want to know that it's from well-fertilized, nutritionally balanced soil.

In some areas pastures are still burned off before planting. Although it's a declining custom, it is justified by its practitioners on the basis of controlling trash and weeds, giving the new growth a better start. Don't do it. Pastures should never be burned. It is far better to turn them over as green manure and reseed. A soil-rich, healthy pasture well seeded will thrive to the point where weeds are a minor problem.

SEEDING

Pastures are usually seeded more heavily than meadows. Regional variations make it hard to define the exact amount of seed to be used per acre; the range is normally from fifteen to thirty pounds. Two good standard mixes are eight pounds alfalfa with ten pounds timothy, or four pounds red clover and three pounds ladino clover with eight pounds timothy per acre. Either of these pasture mixes needs a fair rainfall and well-drained soil, however; on sandy soil in a dry climate it would be a complete failure, except, of course, in irrigated areas, where the alfalfa would thrive.

For meadows, a good basic mix is ten pounds timothy, eight pounds clover, and four pounds redtop per acre. Again, however, this is for good soil and climate conditions.

When you're sowing only one kind of seed, you don't increase the quantity of it to equal the total poundage of a mixture. In the above meadow mix for instance, had you sown only timothy, you would have used fifteen pounds; only clover, eight pounds; only redtop, six pounds.

Pick your pasture and meadow crops on the basis of what will grow in your area. But don't be afraid to experiment. If you can't find a good reason why ladino clover, for instance, shouldn't grow on your land, try some on an extra acre even if no one else is growing it. Farming is a continually changing field where experimentation and new ideas are as important as in any other area. Build on the tried and true, but don't hold back from trying your own ideas once you know what you're doing.

GRAZING

Hay is harvested when the crop is ready—you can tell by looking at it. Your animals, on the other hand, can't tell at all. This is why pastures must be managed. Don't let the livestock graze in early spring until the plants on their pasture are well established and the winter ground moisture has settled. There is a temptation to let the animals out in the fields as soon as they look green. But unless a crop is sturdily on its way before the animals begin munching—this is particularly true for goats—the subsequent summer growth of the plants will be very poor.

For the same reason, don't overgraze a pasture. If the livestock are eating it down to the ground, either expand the pasture or cut down on the number of animals. Pasture should not be grazed lower than three or four inches. Also watch out for the development of a favorite spot where the animals congregate, grazing too heavily, leaving the rest of the pasture to grow unbalanced. Fence off the coffee klatsch corner until its growth has recuperated.

FORAGE LEGUMES

Alfalfa. This is the most important forage crop in the United States. Seeds may be relatively expensive, however, which might militate against using too much of it. Also check with your neighbors about weevil problems in the region before you plant. The three main kinds of alfalfa cultivated are the common, the Turkistan, and the variegated. The differences between them are not worth going into in detail, but they explain why some alfalfa has purple flowers, some yellow, and some a wide range of colors in between, including brown. Stay with the kind they grow in your region, except for maybe a small experimental plot.

The alfalfa plant is a perennial whose multiple leafy branches shoot off from a surface-borne or subsurface crown. There may be anywhere from five to twenty-five branches on a plant, even more under special circumstances. The plant varies in height from a foot and a half to three feet. The roots may be considerably longer, soil penetration of five to ten feet being common. This deep root penetration makes it an excellent crop for improving soil quality.

Most of the alfalfa grown in the United States is found west of the Mississippi because it thrives in a dry, hot climate, as long as the soil itself is moist. However, it's adaptable, and can be cultivated in most regions of the country where deep, well-drained, well-limed soil can be provided for it.

Avoid imported seeds; they fail to yield as well as the domestic varieties. All right, you say, but a seed is a seed. How can I tell a plump olive-green Mexican seed from a plump olive-green California seed? The answer is, you can't. For this reason the Federal Seed Act requires color coding of imported seed. All Canadian alfalfa seed is stained 1 percent violet; it grows quite well, however, in the United States. South American seed is stained 10 percent orange; that from other countries, 10 percent red. Also, naturally, don't buy shriveled or dried-out-looking seeds.

Spring seeding is the most common. The exact time will vary with your locale. It is also common to sow alfalfa with a nurse crop such as oats.

Alfalfa seed, remember, whether planted for forage or as a green manure crop for fertilizer, needs to be inoculated before planting. The seed may be broadcast or drilled. Usually it is covered with one inch of soil, although in the case of heavy soils half an inch is often used, and in sandy soil in drier regions the cover may be one and a half to two inches. Good results have been obtained with broadcasting the seed in early spring on bare, frozen ground, especially fields sown the previous fall with wheat or rye, which again will act as a nurse crop. The heaving of the ground by thawing and freezing covers the seed quite effectively.

Alfalfa pastures, which are excellent forage grounds for hogs, will benefit greatly from a hive or two of bees to fertilize the flowers. And one strong bee colony can make fifty pounds of extra honey from a single acre of alfalfa.

Clover. The best varieties of clover are alsike, ladino, and white. The last two are spreading varieties that form a dense turf. They are thus excellent soil builders, particularly ladino, which has a deep root system. White clover is the most widespread of the group, growing all over the United States. All three are perennial.

Red clover, a good biennial, prefers abundant rainfall and moderate temperatures both summer and winter. For these reasons it is grown primarily in the northeastern and north-central regions, although it will take to almost any part of the United States—including Alaska, for those of you considering homesteading up north. It is important, however, to use locally adapted red clover seed.

Alsike clover, like red clover, is primarily a northern and northeastern crop. It is more cold-resistant than red clover, and will grow on wetter land and in areas where red clover has been grown for a long time but no longer thrives because of "clover failure"—an as yet not fully accounted for condition where the clover refuses to grow even though conditions appear to be favorable.

Ladino, which has white blossoms, does better than most others on soggy, poorly drained soils. It will give you very good pasturage, carrying more livestock than any of the other clovers, but it is sensitive to overgrazing, so keep an eye on the munching level of your livestock. As excellent as it is for pasture, ladino clover makes poor hay because it is difficult to mow and cure well.

Use red and alsike clover for hay, ladino and white clover for pastures. All of them will need inoculation before planting.

But don't let your livestock begin grazing on legume pasture without keeping a sharp lookout for bloating the first few weeks. If the animals get gassy and their bellies distended, remove them from the pasture at once, giving them only dry hay for feed. Then gradually reintroduce them to the pasture once the bloat has subsided. In cases of severe bloating, call the vet at once.

SOWING CLOVER

	Amount to Seed—		Lbs./Acre	Seeding Time	Type of Soil	Soil Cover in Inches	Other
	Broadcast	Drilled	In Mix with Other Crops				
Alsike	8–10	4–5	2–3	Early spring	Any but sand	½–1	Clip first year's growth in August
Ladino	5–10	4–6	2–3	Early spring	Any fairly well-drained, well-watered	½–1	Good grazing lasting 5–7 years, but should not be grazed continuously or lower than 4–5 inches
Red	10–15	6–8	2–4	Early spring to fall	Any well-drained soil not acid	½–1 ½–1	Use locally adapted seed only
White	5–10	4–6	2–3	Early spring to fall	Almost any	½–1	

Lespedeza. This annual or perennial—125 species, two of them annual, the rest perennials—is sometimes called Japanese clover. It comes from eastern Asia, and its shamrock has narrower leaves than clover.

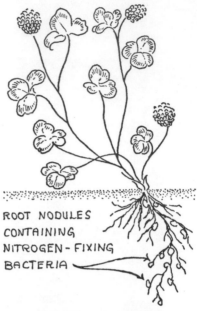

ROOT NODULES
CONTAINING
NITROGEN - FIXING
BACTERIA

ALSIKE CLOVER
GROWS 10-14" HIGH

Lespedeza thrives in hot weather and is killed by frost. It grows on any well-drained soil, reaching a height of from four to twenty inches. The annual varieties are sown in spring, two weeks before the average last frost, the perennials just after the last frost. Unhulled seeds are sown at fifty pounds to the acre; scarified or hulled seeds at the rate of twenty to thirty pounds per acre broadcast. Like all the clovers, it must be inoculated before planting.

Lespedeza hay, when cut in first bloom and field-cured, is almost equal in quality to alfalfa hay. And since alfalfa does not thrive as well in the Southeast, where lespedeza does, it makes an excellent alfalfa substitute for this region. It is usually sown for hay in a mix with bermuda grass or timothy, since without the support of one of these plants it flops on the ground so that you can't cut it properly. In pasture, the annuals are used for summer grazing, the perennials for late fall grazing.

Trefoil. Both bird's-foot and big trefoil are long-lived perennials with moderately leafy stems from a foot to five feet tall. Multiple

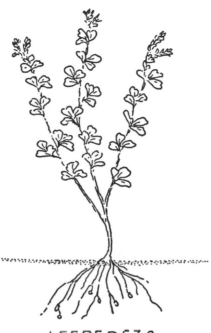

LESPEDEZA
GROWS 4-20" HIGH

stems usually grow from each crown topping the branching taproot. The leaves have five leaflets instead of the three found on clover and alfalfa. Big trefoil generally has more flowers than bird's-foot trefoil, which gets its name from the seed pods angling out from the flower stems like birds' feet.

Trefoil will grow under a wide variety of soil conditions. It is fairly drought-resistant. Bird's-foot trefoil is recommended as a forage crop in the more northern areas of the country, being the winter-hardier of the two. Although trefoil pasture is harder to establish than alfalfa or clover, it has many advantages. For one thing, you can grow it on soil not fertile enough for alfalfa. Secondly, bird's-foot will grow on somewhat alkaline soil, even soil that is clayey, heavy, and poorly drained. Also, once a trefoil pasture is established, it's there to stay, lasting as long as fifteen years without reseeding.

If you're sowing trefoil as a single crop, use between four and six pounds of inoculated seed per acre, the larger quantity for bird's-foot. Usually it is sown with a grass at the rate of two to three pounds of trefoil to six pounds of grass per acre. Bird's-foot can be sown in spring or early fall; big trefoil is usually sown only in spring.

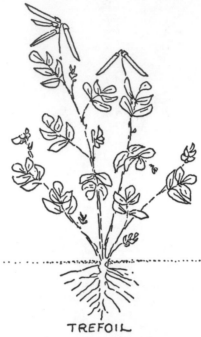

TREFOIL
GROWS 12-24" HIGH

Kudzu. This plant is propagated vegetatively. That is to say, no seeds are used; instead, the crowns that form on the nodes of old plants are set out as new plants in rows twenty to twenty-five feet apart at the rate of five hundred to six hundred plants per acre. For quicker results, cross-rows five by five feet are planted with one crown at each intersection. Kudzu grows only in the southern half of the United States and needs well-drained, permeable soil.

Besides its use for pasture and hay, kudzu is very important in erosion control and soil improvement. As a long-lived, deep-rooted perennial growing well on slopes, the coarse kudzu vine can convert eroding, destructive gullies into prime grazing pasturage in a couple of years. It makes an excellent cover crop on cleared areas.

FORAGE GRASSES

Bermuda Grass. A long-lived perennial that spreads by runners, rootstock, and seeds, it is very vigorous and grows like a weed. The

KUDZU

CREEPING PERENNIAL

original species came from India—you thought maybe Bermuda? Numerous variations of it have developed, among which coastal, tift, and midland are recommended for hay and pasturage. Bermuda grass grows well on most soils, but prefers well-drained, rich bottom land. It has low shade tolerance.

Plant in spring or early summer. If seeding, broadcast or drill at five pounds per acre beneath half an inch of soil. The daily mean temperature should be at least 65°F. before seeding, since bermuda grass is very cold-sensitive. Vegetative cutting is the preferred propagation method. Seven bushels will handle an acre when planted in crosses two by three feet. Cover with about three inches of soil.

You will have no trouble raising bermuda grass in the warmer climates. Note, however, once it is established, you'll have no end of trouble getting rid of it, so don't plant bermuda grass on acreage you expect to devote to field crops in the near future.

Redtop. The most adaptable of all cultivated grasses, redtop will grow in all parts of the country except the extreme South and extremely dry areas. This doesn't necessarily mean that it's an economically sound proposition everywhere, however. If no one in your region is growing redtop, experiment, either using it as part of a meadow mix or by itself in a small stand, because chances are it will do well. But you never want to risk everything on one new crop.

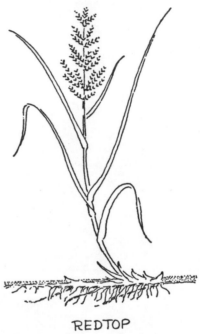

REDTOP
GROWS 12-18" HIGH

Redtop is a creeping perennial with short rootstock and vigorous above-ground growth reaching a height of twelve to eighteen inches. Regionally, it usually matures at the same time as alfalfa. Therefore they are often sown together in a mix for better hay. Variety is the spice of life even for livestock, and hay made from two crops is not only more nourishing than that made from one, but tastier as well.

Seed redtop either in fall or spring. When sown with alfalfa and/or red clover or other legumes, the usual percentage of redtop is four pounds per acre. Seeded alone, eight pounds per acre are used.

Because it doesn't mind slightly acidic soil, redtop is an excellent first-year crop for land that is having its pH raised by liming. You'll get a good healthy crop even while the lime is doing its work.

Timothy. Timothy gets its name from old Timothy Hansen, who introduced it on his farm in Maryland during the early 1720s. It's come a long way since, now being, along with redtop, one of the most commonly grown grasses. However, it is a hay plant, and although second-cutting timothy is sometimes used for grazing, the plant doesn't like it. The two best-known varieties are Marietta and Lorain. (I've never been able to find out which one was Hansen's wife and which was his mistress.) Marietta is the earlier maturing of the two, and as such it is the one grown in more southern regions so it can be harvested before the hot, dry weather sets in. Lorain takes about ten days more to mature and is the one usually used in the North.

Timothy can be seeded either in fall or spring. When seeded alone, which is rarely the case, ten pounds per acre are used; when mixed with red or alsike clover, alfalfa, or other legumes, five pounds are usually used. It does particularly well when grown with clover for quality hay.

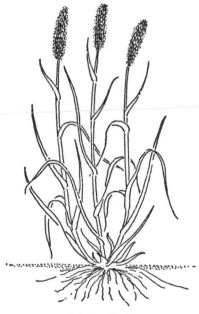

TIMOTHY
GROWS 18-30" HIGH

OTHER FORAGE CROPS

In regions where corn is not an economical field crop, and often in the case of a small farm where silage is not possible, root crops such as turnips, rutabagas, carrots, and sweet potatoes may be used in limited quantities as succulent fodder during the winter. The cost of raising them makes large-scale feeding impractical, and overfeeding of root crops tends to give your livestock diarrhea. But you'll probably be growing them for yourself, and you'll also probably grow more of them than you can eat the first year. Share the remains with your livestock.

MAKING HAY

Over half the farms in the United States make hay. Hay crops rank first or second of all field crops in terms of acreage and value. But, as in all areas of agriculture, hay-making has undergone profound changes. Where once grasses such as timothy, redtop, and bermuda, and even the grains—wheat, oats, and rye—were the dominant constituents of hay, today more and more hay includes in the mixture the legumes—alfalfa, clover, lespedeza, and trefoil. Legume hay has more digestible protein. And more protein means healthier livestock.

Chances are you'll have a neighbor come in and help you cut your hay the first couple of years, in exchange for your help on his fields. It just won't pay you in the beginning to buy the expensive equipment needed. Still, you'll want to know something about when and how to harvest. And if you have only a couple of acres, you may be doing it yourself with a scythe.

HARVESTING TIME

Besides fertile soil, the two major factors determining the quality of hay are its harvesting time and curing. You wouldn't pick strawberries underdeveloped and green, nor would you wait until they became mushy and half-rotted. Hay, too, has its optimum time for harvesting. And, like strawberries, hay can be harvested more than once. After cutting, it will grow back. Then you harvest again.

When to harvest, of course, depends on what you're growing. The following will give you a general idea of what to watch for. When harvesting a combination of legumes and grasses, the time to do your cutting is when the legumes are ready.

LEGUMES

Alfalfa. The leaves of the alfalfa plant contain considerably more protein than the stems, and since the leaves are shed as the plant matures, early harvesting is recommended for maximum feed value. For the highest-quality alfalfa hay, the plants should be cut just when about 10 percent of the buds begin to open into blossoms. At this stage you'll also notice new shoots developing from the crown.

Clover. Red and alsike clover are best cut when in half-bloom. Both will yield a good second crop, under favorable conditions, if this cutting rule is followed.

Cowpeas. This hay is not recommended for beginners because of the uneven ripening of a field of cowpeas. However, for best results, cutting should follow maturation of 10 percent of the pods.

Lespedeza. Cut when in full bloom.

Soybeans. Also not recommended, since the beans aren't ready for harvest until the pods are so heavily laden that the seeds are touching—at which stage the hay quality has already peaked.

Trefoil. Harvest when 75 percent in bloom.

GRASSES

Bermuda. Cut when in full bloom. The hay is most nutritious then, and second growth will also be better.

Redtop. Since it is usually seeded with other crops, such as alfalfa, clover, or timothy, cutting depends on associated plants.

Timothy. The most common hay grass. Harvest in early to full bloom, when it looks dusty blue. The pollen will be loose, and fall when shaken. This is your best bet for hay, particularly if sown interspersed with redtop or clover. In the latter case, harvest with the clover.

If you harvest hay too early, it will be difficult to cure, because the moisture content of the crop will be high. Even when the crop is ready, however, you might have to delay harvesting somewhat if the ground is very wet, since again, the moisture would impede curing. Generally, though, given the choice between hay cut on time but poorly cured and hay cut a bit on the late side but well cured, hay cut a little late is preferable. Hay cut too early, besides not curing well, retains less nutritional value and volume.

On the other hand, cutting hay too late makes waste. For one thing, if it is too dry, particularly in the case of the legumes, the leaves will have dried out to the point where they fall off during harvesting and handling. Secondly, hay that has overmatured will tend to lodge. That is to say, it will dip down to the ground, particularly after heavy rains. This makes it difficult to machine-cut, and a large part of your crop gets left behind in the field.

CURING THE HAY

The more quickly hay is cured, or dried, the better its quality. This is because curing is actually more than just drying. During the dehydration process a certain amount of fermentation occurs. This fermentation is induced by enzymatic action, which makes the hay more digestible. It also adds aroma and flavor. Too much fermentation, on the other hand, produced by too much moisture over too long a period of time, will heat the hay up, reducing its quality.

Hay freshly cut is about 75 percent water. Curing brings the moisture level down to around 20 percent. By having the hay crushed as it is cut, you can halve its curing time. The usual practice is to let the hay dry where it lies. Then it is raked into long windrows where it continues to dry, but more slowly, reducing the water content by another 5 to 10 percent. Hay at 20 percent moisture is safe for baling. Hay at 15 percent moisture is safe for storage.

But how do you tell what the moisture content is? Pick up a handful and make a bundle of it so the stems are parallel. Wring out the hay as you would a wet cloth, then bend it into a U. If your bundle is rather wilted-looking and breaks too easily when twisted and bent, it's still too wet. If it's pliable, like a piece of rope, and takes a lot of twisting and bending to break, it's just right. If it's crisp like celery, snapping off clean just from being bent, it's too dry. But all is not lost. If your hay is too dry, just wait till the next morning. The dew will dampen it again.

You'll get the feel of hay by talking with your neighbor farmers and checking out their crops. There's no substitute for actually handling and smelling the hay. And it's important to know your hay. Poor hay is not only bad for your livestock, with moldy hay capable of killing them, but poorly cured hay has started many a fire. Hay stored too wet will ignite by spontaneous combustion, and you'll lose not only your crop, but barn and livestock as well.

THE HAYSTACK

Most hay today is stored in bales. There's less waste that way, it's easier to handle, and since the hay is tightly bound, it takes up less room in the loft. If someone cuts your hay for you, chances are he'll bale it as well. If he doesn't bale his, or if you cut your own, storing it loose will still work fine. Chances are you won't be raising enough just for your own use to fill a barn anyhow. But remember to get the hay off the field quickly and under cover as soon as possible. This is doubly important with loose hay. Outdoor haystacks are picturesque, but not very good farming, except in the dry areas of the West.

SILAGE

Rain has ruined many a hay crop, but it doesn't affect silage, which helps to explain partially the popularity of this method of storing winter fodder. If your livestock population is small, silage may not be a practical venture for you. But it's still well worthwhile knowing some of the basics, both because they're interesting and because you never know how rapidly your herd of farm animals might expand. If you have a barn and if it comes with a silo in fairly good condition, you might begin using it sooner than you expect.

Although some people interpret the ancient Roman grain-storage structures as having been suitable for silage, chances are the actual ensiling process is of German origin dating from the mid-1850s. The first American silo was built in 1875. Considering man's lengthy involvement with agriculture, that makes silage a very new development.

By far the most common crop ensiled is corn, followed by the grasses and the legumes. In some of the northern sections of the country, as well as in the western mountain regions, sunflowers, because of their frost resistance, are popular for silage.

A crop harvested for silage, unlike hay, must be chopped up before storage. This involves a machine called a forage harvester. Ensiling isn't exactly on a par with chopping cabbage for sauerkraut, and a forage harvester is expensive. Again, you'll probably find a neighbor willing to use his machinery on your field in exchange for your help on his. He'll also be able to tell you the best time to cut. Plants for silage are usually not permitted as much growth as those for hay.

A general timetable for cutting some of the more popular silage crops is as follows:

Alfalfa. Cut when about 25 percent into bloom.

Clover. (Red and Alsike). Cut when 50 percent in bloom.

Corn. Cut when the early ears are well ripened and the rest are beginning to turn.

Lespedeza. Cut when the first seeds begin to form.

Sunflowers. Cut when the petals are dry and beginning to fall.

THE SILO

The silo itself can be one of four types. The big round tower usually seen in typical-farm-scene-calendar shots, and from which the missile launch protective system derives its name, is the traditional model. Commonly constructed of wood staves like a barrel, it's the kind that's apt to come with your barn. Newer models are also made from brick, poured concrete, sheet metal, or glass-lined steel plate.

The pit or trench silo is just what the name implies—a long horizontal trench into which the silage is poured. Very economical to construct, it nevertheless has several drawbacks, not the least of which are the labor involved in getting at the silage and the high silage loss of up to 20 percent, as compared with around 5 to 10 percent for the tower type.

Bunker silos are large above-ground boxes into which the silage is loaded. Although cheaper to construct than the tower variety, the silage loss used to be too great to make them a viable economic proposition. However, by lining the box with sheets of polyvinyl chloride plastic and capping the silage with more of the same, losses can be kept low enough to compete with the tower in efficiency.

The fourth type of silo is more a method than an actual physical structure. Silage is simply piled up directly onto a concrete slab, or even the bare ground. The rotting of the outer layer forms an airtight cover around the stack, acting as its own ensiling structure. This method is relatively laborless and cheap. Unfortunately, surface losses are so great as to make it a much more expensive process than it would at first appear. With the addition of a plastic bottom sheet and cover, however, the picture changes. High-quality silage is produced with minimum seepage loss. Also, the plastic film method permits making small quantities of silage of as good quality as you'll find in a fully filled tower silo.

FERMENTATION

But what exactly happens in a silo? Is it just a fancy way of storing shredded hay? Not at all. Silage is to hay what wine is to grape juice. Fermentation is the key in both cases. The first change that occurs when freshly chopped, moist silage is placed in the silo is conversion of the oxygen into carbon dioxide. Within two or three days almost 70 percent of the silo's total volume of gas is carbon dioxide. After peaking at this point, the carbon dioxide content will begin to decrease again. The second change is that for the first two weeks the temperature of the silage, like that of your compost heap, rises slowly. Toward the top of a tower silo it will reach well over 100°F. by the end of the first couple of weeks.

Both of these changes are due to acid fermentation within an airtight structure. Ethyl alcohol, acetic acid, and lactic acid are produced by the bacteria thriving in silage. However, as these ingredients increase, the bacteria count decreases rapidly. Silage then is almost like a self-canning process. Once the bacteria have disappeared, the silage will keep indefinitely, as long as no fresh air is allowed to enter into the closed system. If any air does enter, mold will form rapidly, spoiling the works.

When fermenting beer, sugar is often added. Although this is not a necessary step in silage, it is often a helpful one for preservation. It also cuts down on odor—and there's some silage you'd otherwise definitely want to remain upwind of. The sweetener used for silage is molasses, which contains 50 percent sugar. The molasses is mixed in as the material is cut, or, in the case of tower silos, when it is blown to the top of the silo.

In addition to the molasses, when making grass and legume silage ground barley, corn, oats, wheat, and even corncob are often added. These supply both sugar and starch. Starch in fermenting increases the acidity of the silage, which in turn increases the keeping quality of what is then, appropriately enough, called acidified silage.

The wilting method of preparing silage requires no preservatives. Just let the legumes, grasses, or small-grain crops field-dry, less than you would for hay, however; a moisture content of about 60 percent is what you want. The vegetable matter must still be chopped as with other silage. In addition, the drier wilt silage must be packed more tightly than the preserved variety. To see if the crop is ready for your silo, squeeze the chopped "hay" tightly in your hand. Open your hand. If the ball is juicy, it's still too wet. If it isn't juicy, keeps its shape or falls apart slowly, and your hand is moist, it's just right. If the ball springs open, you blew it; it's too dry for ensiling. Let a morning's dew moisten it down and try again.

In modern glass silos the bottom layer of silage, which has aged the most, is used for feeding. The glass silo acts like an endless sausage casing, raw material in the top, finished product out the bottom. The older silos work on the last-in first-out principle. Remember that if silage is well packed, success is almost guaranteed. You'll never catch anyone making love in the shredded hay and molasses of silage instead of in the hay loft, but for good fodder production the silo is hard to top.

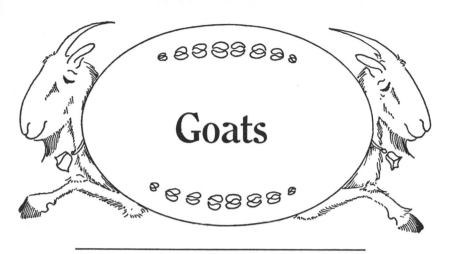

Goats

God gives the milk but not the pail.
—OLD ENGLISH PROVERB

MILK AND CHEESE are staples in our diet, and the apprentice farmer's first thoughts on dairy matters are apt to be of a cow. But while there's nothing wrong with having a cow on your spread, sometimes she's more trouble than she's worth.

A cow will produce twenty quarts of milk or so a day, which is a lot for one family to consume, yet not enough to make a real business proposition out of daily milk sales. In a communal situation, of course, consumption would be no problem. But milking might. A cow will adjust to being milked twice a day at just about any hour as long as the schedule is regular; however, milk production is best if she deals consistently with the same person. Unless you have a real bovine buff in your midst, this might present a problem in labor allocation.

It may well be simpler to build up your flock of hens to the point where you can swap eggs for milk on a regular basis with a neighboring dairy farmer. Still, by the time you've been in the country a year or so, you may want a milk-and-cheese source right on your property— maybe meat and leather too. Well then, a goat is your answer.

There are many advantages to goats, not the least of which is that goat's milk, sweet and tasty, is more readily digestible than cow's milk, and hence more nourishing. You are what you digest, not what you eat. Also, goats don't get tuberculosis, so you don't have to worry about

192 ᵫ GROW IT!

pasteurization, which not only takes time and equipment, but, as with any other food-heating process, can cause vitamin loss as well.

Goats are basically good-natured (remember, however, that the word "capricious" comes from the Latin *caprus,* meaning "goat"). They are easy to care for, clean, and excellent fertilizer producers. Goat manure ranks far above cow manure for nitrogen content and is surpassed in this important fertilizer component only by rabbit and chicken manure, placing it high on the list of desirable droppings. So why not a goat or two?

BREEDS

Goats, by their nature, will fare well in any part of the United States. This goes for any of the different breeds, and so selection is made on the basis of other factors. One fairly important for the apprentice farmer is the preferences of other goat raisers in the area, if there are any. If people in your vicinity are raising Alpines, don't raise Nubians. Not that the other goat breeders would refuse to help you, but they are a bit clannish. Besides, you're better off getting your goats from someone you can talk to later about any problems you have. A good goat's a good goat, and a bad one's a "bad 'un," no matter what her breed.

Whichever breed you do settle on, stay with it. Don't try raising two or three different kinds at once. Familiarity breeds knowledge, and as a beginner you don't want to spread yourself too thin.

In goat trader's lingo, a "pure breed" is just what its name implies: both parents are known and of the same registered breed. A "grade," on the other hand, has a registered purebred sire, with the other parent of unknown or mixed pedigree. The latter is cheaper, of course, and will serve your milk pail and cheese hoop quite adequately. On the other hand, if you plan to stay on the farm you might consider buying purebred registered goats and breeding your own herd. Part of the joy of country living is helping to bring out the best in Mother Nature.

Nubians. These Roman-nosed beauties are of English-African-Oriental descent. They have a short, sleek coat ranging through a broad variety of colors and color combinations. Their ears are long and drooping.

Saanens. Among the most prolific producers of milk, the white/cream-colored Saanens originated in the Swiss valley of the same name. They are large, relatively short-legged, straight- or concave-faced, with small, erect ears. Some purebred Saanen does have produced over five thousand pounds of milk a year.

Toggenburgs. These are the most popular goats today—at least to judge by nose count. They fit the common image of a goat, brown, with lighter markings down the face and inside the legs joining a light-colored underbelly. They have erect ears, a gently concave face, and an extended tail. Native to the Toggenburg Valley of Switzerland, they were the first goats imported to the United States in quantity, which accounts for their numerical strength.

French Alpines. Color varies from pure white to spattered black. The body configuration is similar to that of the Toggenburg and the Saanen, although the eyes are more prominent.

Whatever the breed, a goat will have its basic good points and flaws. If you buy locally, try to get one that's between two and four months old, freshly weaned. Inspect the animal before purchasing, checking that the goat has a straight back line. A back humped between the fore and hind legs is known as a roach back, one with a depression as sway back. Avoid both. The rump should be long and not slanted down toward the tail too steeply. The chest should be wide and the heart girth, the circumference around the front ribs, should be large. Basically, you want a goat that is rugged looking, with straight legs, standing squarely on its hoofs. Avoid goats with horns and you'll avoid a lot of hazardous ramming around the barnyard. If possible, have a look at several different goats so you can get a natural feel for a good, round, symmetrical udder (although rarely are both halves exactly alike) with well-located teats pointed slightly forward.

I'd rather drive two hundred miles to look over and pick up my goat than have it shipped. But if you have to order by mail, ask the breeder for bank references and the names of previous buyers. Insist on a written contract and a veterinarian's certificate stating the animal is free from disease and parasites.

Once you've located a reliable breeder—ferret him out through the *Dairy Goat Journal* or your area farm paper if local advice doesn't

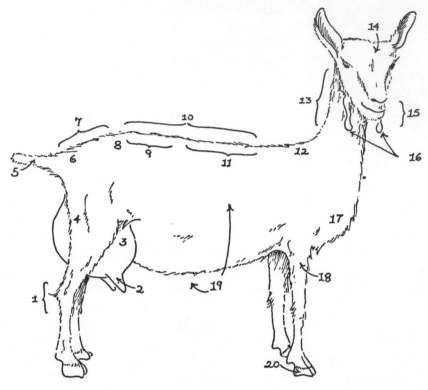

MILCH GOAT

1. HOCK	6. TAIL HEAD	11. CHINE	16. WATTLES
2. TEAT	7. RUMP	12. SHOULDER	17. CHEST
3. UDDER	8. HIP	13. NECK	18. FOREARM
4. THIGH	9. LOIN	14. FOREHEAD	19. BARREL
5. TAIL	10. BACK	15. MUZZLE	20. HOOF

unearth one—you might as well follow his recommendations in pur-chasing stock. But remember, a breeder rarely sells his best stock. He's spent years building up a breeding herd and isn't about to skim off the cream of his labors for you. This doesn't mean he's going to sell you second-rate stock either, of course. Just don't be disappointed if your doe's milk-production record is not quite up to that of her mother. And a goat giving four quarts a day for a professional might produce only two quarts for the amateur—but you don't stay an amateur forever. Another consideration is that goats get terribly homesick and will often produce less their first year in new quarters, particularly if alone. Keep-ing two goats, if you can, lessens the problem greatly.

HOUSING

Keeping goats is one of the simplest livestock operations. The basic need is for a dry, well-ventilated but draft-free shelter that is easy to keep clean. You can remodel a garage or outbuilding for your goat stable, or you can construct one from scratch. The space requirement for stanchioned goats is about three by five feet per animal plus two feet for a feed alley (that's for you to walk along while tending the animals). If the goats are to be housed in a barn without stanchions, as is recommended, plan on ten to twelve square feet per head.

For cleanliness, a gently sloped concrete floor with good drainage must be provided. In severely cold climates a raised wooden platform, one which can be moved for cleaning, should cover part of the cement so the goats will have a warm spot to sleep. Even the thickest bedding might not insulate the cold concrete enough in really wintry weather, and, besides, the thicker the bedding, the more difficult the cleaning.

If possible, the goat barn should be located near a source of water pressure sufficient for it to be hosed down once a day. But be sure to let it dry thoroughly after hosing before you allow the goats back in. The barn should also be disinfected occasionally with a good scrubbing, using ammonia or pine tar and water. Nicotine sulphate is highly poisonous, to man as well as to insects and bacteria, and is not recommended as a disinfectant, nor is creosote spray, which is also readily absorbed by the milk.

Painting the inside of the barn as well as the outside is a good measure for disease prevention and highly recommended. Avoid lead-based paints, since goats may nibble on flaking surfaces and die from lead poisoning. Old-fashioned whitewash, homemade without DDT, is still best.

WHITEWASH

1 oz. casein glue

10 oz. hydrated lime

Make a separate, smooth paste of each, using water. Stir them together with enough water to make a slurry. Mix well. Add:

1 teaspoon salt

Thin with more water. The final wash should be thin enough so that when brushed onto a newspaper you can still read the small type. Use two coats or more.

Ventilation is essential in a goat barn, as in all other livestock housing. One of the big variables here is ceiling height. As one moves from the southern United States to the North, ceiling height should decrease proportionately in order to keep in the heat, and hence total floor space might have to be increased. Plan on having one hundred cubic feet of air space per goat and at least one square foot of openable window for each two hundred and fifty cubic feet of barn space. Even more window space is better, since sunlight is a natural disinfectant. However, if you're located where the winters tend to be severe, make certain the extra windows can be covered by a double pane or shutter in the wintertime so the inside temperature can be regulated to around 55°F. Too cold, and the goats will concentrate on keeping warm instead of making milk. Don't make the opposite mistake of trying to keep the indoors temperature in the pleasant 70s; the goats, overprotected, will chill when they go outside.

Year round, all windows left open should be screened to minimize flies. The barn doors, including the one you use, should be solid, not screened. If you use a screen door, the flies will hang around just waiting for you to open it, and they'll beat you inside every time.

In the summertime the barn is an important source of shade. Goats are not bothered by heat as long as they can avoid too much direct, prolonged sunlight, a matter about which they know best, and are assured an ample supply of clean, fresh drinking water.

EQUIPMENT

The goat, as opposed to, say, a chicken or a horse, is a wasteful feeder. Food spilled on the floor will not be eaten. Therefore, well-designed mangers and grain feeders are important to an economical goat operation.

Water can be provided in a bucket or trough, anchored so the goats will not knock it over. The disadvantage of this form of watering is that goats are quite fussy as to what they will drink. And drink they must, in quantity, to produce milk. Even so, they won't touch contaminated water.

For this reason I prefer a running water system wherever possible. An old sink, or a specially constructed raised trough with a bottom drain and a lead-off pipe, can be used where the water supply permits continuous filling. It doesn't have to be a great rush of fresh water

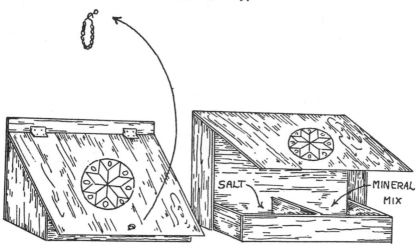

GRAIN FEEDER WITH CLOSING TOP   MINERAL FEEDER
TO PREVENT CONTINUOUS FEEDING   OPEN FOR CONTINUOUS ACCESS

MAKE COVER STEEP ENOUGH SO GOAT   ROOF KEEPS GOAT FROM CLIMBING
CAN'T JUMP UP AND STAND ON IT.   INTO FEEDER.

every minute; a steady trickle, pulling contamination down the drain, will keep your goats happy. If your water source is very cold, you'll have to build a holding tank inside the barn to let it warm up before it reaches the trough.

To go with the water you will want a salt lick. It coaxes the goats to drink more, at the same time supplying a necessary part of their diet.

Toys are essential in bringing up goats. A milking doe should not be allowed to jump about too much, since the energy so expended will cut into milk production. However, dry and growing stock need plenty of exercise. If your barnyard isn't naturally rocky, make some mountains—a sturdy crate or two, of different sizes, for them to jump on, an exercise ramp and—don't laugh—a seesaw. Bucks like to play with an empty barrel or a swinging tire.

The primary objective of having a dairy goat is, of course, milk. The equipment needed here can range from basic to fancy, but in all cases will include a screened-off room or separate milk house—a separate house is preferable—to aid in keeping out flies. (If you get the idea that flies are one of your biggest dairy problems, particularly in the summertime, you're right. Keep them under control with a clean barn, and remember to always close the door behind you.) In your milking shed you'll need a milk pail, a strainer, storage containers (bottles and milk cans), and cooling facilities, either electric or ice buckets. In small bottles to insure quick cooling, fresh milk for family consumption can be placed directly in the refrigerator next to the cooling unit. In any case, it must be cooled to 45°F. as quickly as possible. Milk held at room temperature will deteriorate rapidly. One last item: A milking stand is helpful mostly to you, since it raises the udder to a level where you can milk comfortably while sitting on a stool. Any low platform for the goat to stand on will suffice, but it should have a stanchion at one end to keep the goat steady and in place.

FEEDING

For the novice farmer in particular, the best nutritional results are probably produced by using a balanced commercial goat feed, carefully following the manufacturer's instructions. If you want to blend your own mix, ask your county agent for advice. He'll be listed in the phone book under "County Extension Service." Should you happen to be in an area relatively devoid of goat farms, he might say he's not familiar enough with the animal to help you. Don't let that be a deterrent. Goats will do well on the same mix he recommends for milch cows.

GRAIN MIX FOR GOATS

40 lbs. ground corn
35 lbs. ground oats
10 lbs. linseed oil meal
12 lbs. wheat bran
1½ lbs. salt
1½ lbs. trace minerals (prepackaged)

Goats are as choosy in their feeding habits as in their drinking. Not only do they not eat tin cans, they don't even like food that has been nosed over in their feeding troughs all day. Less food, more often, is the standard rule-of-thumb. A minimum of two feedings ten hours apart is recommended; three evenly spaced feedings of less quantity are even better. A lactating goat, eating all the grass and hay she wants, will have a well-balanced diet with the addition of a total of four to eight pounds of grain a day. Here are some general guidelines to goat feeding:

1. Always feed your herd normal rations before sending them out to pasture. This is particularly important in spring when they are switching over to fresh greens. Allowing them to fill up on all those green goodies after the long wintering can cause severe bloating.

2. It is usually economically handy to switch from mix to mix depending on the season, but don't do it overnight. Change the diet gradually over the period of a week.

3. Consider grain an addition to the normal feed ration, not its main component. A goat's stomach is designed to handle bulk from grazing, so good-quality, well-cured alfalfa, clover, or soybean hay should be the basic staple of the animal's diet.

4. If hay is scarce and you must increase grain feeding, make sure the mix has a generous helping of wheat bran or dried vegetable pulp to add bulk. If her diet isn't bulky enough, your goat will have digestive problems, and in turn less milk to give. Lack of appetite and lighter-than-usual, crumbly manure are the first indications of insufficient bulk in the feed.

5. Whenever possible, let the goats graze naturally in the pasture. The animal prefers a legume pasture to a grassy one, however. Prepare for this by sowing alfalfa, clover, and the like. Also, goats are browsers; that is, they prefer eating somewhere between the ground level of cattle and the treetop level of giraffes. There should be plenty of bushes and shrubs in their pasture. Some people try to take advantage of the browsing pattern by pasturing goats on wild, uncultivated land that is to be cleared, so they will eat it clean; this practice will not provide the goats with an adequate diet, however, unless well supplemented.

6. Milking does must, as all other livestock, have continuous access to salt. Provide a salt or mineral lick. It is particularly important that the salt for freshened does be iodized to prevent goiters in the young kids.

7. Calcium and phosphorus are essential to the feed mix. These are best supplied in a block, or from a mineral feeder.

8. Keep feed clean at all times. This includes keeping other stock, such as chickens and even cats and dogs, out of the hay.

9. Remember there are three types of feeding: the grain mix, which is a supplement to balance the diet and not to be overused, manger feeding of hay, and fresh forage. The last two should constitute the largest part of your goat's diet.

SIMPLE HALTER FOR LEADING GOAT

COLLAR AND LIGHT CHAIN FOR KEEPING GOAT TETHERED IN OPEN FIELD

BREEDING

It takes two to get milk. Even the best-bred doe will not produce unless she is mated and kids. You may choose to buy a doe that is already freshened, that is, she has given birth and is lactating. Or you may want to start with a young doe and raise her to the milking stage. A young doe can be bred to freshen at fifteen months. The best breeding period is fall through early winter.

Starting with a young doe is recommended for the beginner. The initial lack of milk will be more than compensated for by the opportunity the young doe provides for the new farmer to gradually learn about goats and how they grow. Besides, the farm will boast at least one new goat—and usually two or three—once the young doe is bred. On the negative side of the ledger, it means stepping right into five months of prenatal care, delivery, and a month of aftercare rather early in your dairy career. But there's an amazing pleasure and excitement to be found in watching the kid grow day by day. And you'll have to learn about kid-sitting sooner or later anyway. Even a once-freshened doe won't give milk indefinitely, so you'll want more than one, and you might as well have the joy of raising your own.

Whichever way you choose to start your goat herd, you're going to need the stud services of a buck eventually. At first glance, a buck of your own would seem the most direct approach to the problem. But think twice about the idea. It is really not economical for a small-scale operation. As opposed to a doe, a buck is smelly even in the best of conditions. Hence he must be kept separate from your does to avoid odor contamination of the milk. His services are needed only once a year per doe, but he eats all year round. And he's a lot more temperamental and active than a doe.

A better alternative is to hire a buck from a nearby farm as a stud (another good reason for raising the same breed your neighbors do). Stud service can also be handled by mail, the animals being shipped back and forth by express. A third alternative is to form a cooperative association of new goat raisers, sharing the costs and responsibilities of keeping the buck.

KIDDING

Keep a record of when your doe was bred. The gestation period for goats is 145 to 155 days, or roughly five months. About the 140th day, your doe will begin to become restless. She'll seem nervous and a little irritable. She will probably paw her bedding and bleat quietly and often. Have a draft-free, heavily bedded private stall ready for the kidding. As the actual birth draws nearer, she will lie down and get up repeatedly. At last she'll stay down and begin labor.

For the first time, have someone familiar with the process around. You'd probably have no trouble on your own—after all, goats were giving birth long before man domesticated them and your doe's instincts will carry her through. But if you've never been around kidding before, you probably wouldn't instinctively recognize the signs of any complication or how to respond. And the object of the game is to raise a kid or two successfully and to learn in the process.

KIDS AND WEANING

The usual practice in kidding is to take the kid from the doe as soon as the proud dam has cleaned it off. Be sure not to let the kid nurse or you may have great difficulty coaxing it to pan-feed later. Wipe the kid dry with clean cloths and place it in a well-bedded box, preferably in a sunny spot in the barn. The sun will help dry and warm the kid and act as a mild natural disinfectant. Sunshine is just plain healthy.

A newborn kid is sensitive to abrupt temperature changes. Since kids in nature are born outdoors, their resistance to cold is greater than you would expect, and kidding can be done outdoors, but if the new-born kid is then taken into the barn, it must be allowed to adjust gradually to the temperature change. The same holds true when taking the kid from the barn outdoors.

Some breeders prefer to let the kid nurse from the dam, or mother doe, the first couple of times. *Don't do it.* Milk the dam and then feed the colostrum-rich first milk to the kid. The colostrum, thick and yellow, is full of vitamins needed by the new kid.

A new kid will want to be fed soon, usually within the first six hours. You'll know when from the look on his face and the way he moves. At such time, milk the dam. The kid may be either pan-fed or bottle-fed. Bottle-feeding requires a little bit more work; particularly, more things to sterilize. On the other hand, it ensures that the kid will not drink too fast and is a lot more fun. Don't let the milk cool down before feeding. Should there be a delay, you will have to reheat it to about 105°F., or slightly above body temperature. Feeding cold milk produces scours, or acute diarrhea.

Newborn kids should be fed about one-third of a pint five times daily the first two days, a half a pint four times daily the next twelve days. Feed more if the kid is constantly hungry, less if diarrhea develops. Vary the amount fed rather than the number of feedings. At the end of the two-week period, feeding can be reduced to three times daily, while at the same time the quantity is increased. Remember, however, overfeeding is more harmful than underfeeding. Also, to ensure good growth and to flush out the kid's system, let·it drink as much warm water as it wants five minutes after each feeding. The addition of half a teaspoon of cod liver oil a day to the diet will help the kid along.

You probably won't want to use all of the dam's milk to feed the kid or kids. Either powdered skim milk or Dairyaid (a specially prepared formula) is a perfectly acceptable substitute after the first week. Mix according to directions and ease the kids onto their new diet slowly. Second-week feeding should be one-quarter skim milk, three-quarters dam's; third week, half skim, half dam's, and so on. However, be sure the initial feedings use only the milk the dam gives, since the colostrum of the first milk is essential to the kid's well-being.

A kid is usually weaned by the time it is six weeks old. Lead up to this starting the fifteenth day by adding a couple of tablespoons of finely ground cooked oatmeal with a pinch of salt to each feeding. (By this stage, of course, you will have to enlarge the hole in the nipple for the oatmeal if you're bottle-feeding.) Increase the oatmeal each day until by the fifth week you are feeding half gruel and half milk.

Sometime in the second week the kid will probably start nibbling at a little choice hay. Start supplementing its diet then with some kid-starting ration. Leave a small quantity around at all times. Commercial kid rations, the result of professional testing, are highly recommended for the apprentice farmer. You'll have so much to learn and get a feel for with your first kids that you might as well take advantage of this "instant farmhand." Follow the instructions and you'll end up with a healthier goat than Mother Nature's sometimes erratic rations in the wild could produce.

Remember, cleanliness is always important, but when it comes to young farm animals, cleanliness is imperative. A kid's accommodations must be kept always dry, sheltered from drafts, yet well ventilated.

Exercise and sunshine are essential. Let the kid out to play when he's frisky.

THE DAM

For a few days after kidding the fresh doe should be fed lightly. Too much food will raise her milk output too rapidly and may cause mastitis or other udder troubles that could permanently impair her milk production. Check to see that the udder does not become overfilled and hot. If it does, milk the doe three or four times a day, or as often as needed to relieve this condition. Within three or four days the first rush of milk should slow down and the signs of colostrum should have vanished. Even so, the doe will not reach peak production for about a month.

MILKING

You can lead a goat to the pail, but she isn't going to milk herself. Nor are you by merely pulling the teats randomly. Milking isn't all that difficult, and once you've gotten the technique down pat, it will be automatic. The best way to learn is to watch a local farmer. If for some reason there's no one around to teach you, a good way to practice before tackling the goat herself is with the Richard W. Langer nonpatented artificial udder. Take two long balloons. Pinprick holes in their tips. Attach each one to a bottle half-filled with water. Tape the bottles together, invert, and go at it. It's not the real thing, but if you can milk balloons well, you'll upset the goat less when it's her turn.

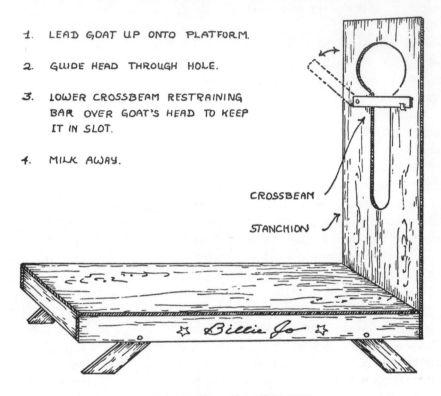

1. LEAD GOAT UP ONTO PLATFORM.

2. GUIDE HEAD THROUGH HOLE.

3. LOWER CROSSBEAM RESTRAINING BAR OVER GOAT'S HEAD TO KEEP IT IN SLOT.

4. MILK AWAY.

CROSSBEAM

STANCHION

MILKING PLATFORM

The goat should be led up onto a milking platform with a stanchion. Milk readily absorbs odors, so pick a milking spot well removed from all possible sources of olfactory contamination. The stall, or preferably a separate shed, should be whitewashed, screened against flies, and free from dampness; in general, it should approach the sanitary conditions of a hospital.

Once the goat has mounted the stand, something she will do happily with a full udder once she's used to being milked there, place her head in the stanchion.

Cleanliness is paramount. The doe's hindquarters and udder should be kept trimmed (more on that later) to avoid hair contamination of milk. Before milking, sponge down the udder with a clean cloth and warm water, then dry.

Place a stainless steel bucket, presterilized with boiling water, beneath the udder and take a seat at the goat's side—it doesn't matter which, but always use the same side. Goats like the comfort of consistency. For the same reason, it's best that the same person always do the milking. (Although in some areas goats are milked from the rear, this is a practice to be avoided, since it causes sanitation problems and has no real advantages.)

You're ready to go. Take the furthest teat in the V of your thumb and forefinger, near the udder. Close the two fingers in a scissorlike motion to prevent the milk from being forced upwards. With a slight downward pull but a firm grasp, squeeze in the second, third, and little finger in sequence. Release. Now the other teat with the other hand. Release and start again. You'll have to develop a touch for it, because all anyone else can tell you is just not to squeeze too hard, yet not too gently either. The single most important factor is to make your motions smooth.

Milk both teats, alternating each stroke, until they are empty. The udder should always be completely drained. When the milk ceases to flow, give the udder a light massage by the teats and "strip" the udder by passing the hand down each teat as before, but without releasing any of the fingers, thus draining the last drop.

MILK PROCESSING

Anything to do with milk processing must be absolutely clean. All utensils must be sterilized—with live steam where possible, boiling water where not. You can get a small steam generator, which is really nothing more than an oversized teakettle, quite inexpensively.

Use seamless stainless steel containers wherever possible for the actual milk processing. Glass bottles are ideal for storage. Rinse them well, follow with a soap-and-warm-water wash and yet another hot-water rinse, then scald with live steam if possible. Boiling water is a barely acceptable substitute. *Do not wipe dry.* Allow to air dry before using.

The milk should be strained as soon as you've finished milking. Disposable sterile cloth strainers are very good for this, cutting down the amount of cleaning necessary. Don't expose the milk to direct sunlight for a prolonged period, since this may cause excessive oxidation and bad flavor.

Rapid cooling of the milk is essential. After straining, bring its temperature down to 45°F., or slightly less, by placing the containers in buckets of iced water if you haven't invested in an electric milk cooler. Use a thermometer to make sure the desired temperature is reached and held for thirty minutes. Small batches such as you would get from one or two does can be placed directly into the refrigerator. However, if you do this, make sure you place the milk in prechilled bottles no larger than quart size as close to the freezer unit as possible to insure rapid cooling.

MILK PROBLEMS

If you've spent all your life drinking milk out of paper cartons, you may be surprised to learn that a lot of things can spoil milk between the udder and the cup. Some things even before it leaves the udder. A goat can be thought of as a natural milk machine. As such, the end product is no better than the raw materials used, that is to say, the food eaten.

Food flavorings—good and bad, from clover to onions—that affect milk are usually caused by volatile plant oils just as are many more familiar food flavorings—from vanilla to cloves. These oils are absorbed into the bloodstream and, because the vital udder draws heavily on the circulating blood supply, their flavors are easily absorbed by the milk. So unless you really like onion-flavored milk, don't let your does graze in pastures overgrown with wild onions. Turnips, cabbage, and potato greens can also impart a bad taste to milk. Garlic is absolutely fatal to the flavor. Grains, on the other hand, tend to sweeten the milk, and thus are desirable as feed from this as well as from a nutritional standpoint.

There are other plants you should keep from your goats. Not because they will directly produce off-flavored milk, but because they may make her ill, or even cause death. The best way to learn what dangerous plants might affect your livestock in any particular region is to talk with the locals. As a start, goats should not be allowed to graze on azaleas, laurel, milkweed, and wild cherry.

Your doe's health will also affect the quality and flavor of her milk. A goat that isn't feeling up to par will still produce milk—after all, if nature had not so prepared it, a wild goat herd might have become extinct after a severe epidemic of winter colds—but that doesn't mean milk from sick goats is desirable. A herd or a single goat should always be kept in the best physical shape possible.

Blood in the milk might not give it a bad flavor; nevertheless it's an obvious sign of trouble. It indicates possible injury to the udder, such as bruising, an imbalanced diet, or mastitis. Although you will be able to handle these problems by yourself once you know how to diagnose them, it is best to get either your veterinarian or a local dairyman (milch cow men tend to scorn goat-keepers, so get an acquaintanceship going before trouble arrives) to diagnose the problem for you and discuss the solution.

You may be talking to the veterinarian more than you expected to the first year, but information like this really can be transmitted only in the old oral-visual tradition. When contacting a veterinarian, make sure that he specializes in farm animals, and is not a suburbanite poodle-and-pussy doctor. The pet vet will probably be just as competent, but your bills in all likelihood will be three times as large. In contrast, I know of one local farm veterinarian who takes care of a goat herd in exchange for a wheel of cheese now and then.

If you learn to recognize symptoms of a goat's lack of the best health, you're already doing well. The rest will come with time, and talk, and just watching.

After fresh milk has stood a few hours, you may notice a redness to it that you didn't observe while milking. Chances are it's a bacterial by-product rather than blood. Poor sterilization is the usual cause, although sometimes failure to cool the milk rapidly enough will also induce the same undesirable tinge. The remedy is greater cleanliness.

Any utensils that have come in contact with infected milk must be extra carefully washed and sterilized. It's also a good idea to wash the doe's udder thoroughly.

Ropy milk, or "long milk" as it's called in Sweden, is an ancient favorite of some north European farmers. However, even there its popularity is fading rapidly, and in the United States it has never been appreciated at all. Politely speaking, the milk becomes slimy and stringy, although where I grew up we called it snotty. This change in its physical characteristics is also caused by bacteria, harmless enough ones, either in the udder or after milking. Here again, the problem usually can be solved by increased sanitation.

THE DRY DOE

A young doe may be freshened for the first time at about fifteen months if she has been well raised and is in good health. However, she's not going to produce milk for the rest of her life from one freshening. Trying to stretch out her milk-productive period too long without freshening again will impair her health and generally cause nothing but trouble. A goat should be freshened once a year. So to ensure a steady supply of milk it's advisable to have two goats and freshen them alternately.

Before freshening a doe that has been giving milk, she must be "dried off." The method used by most dairymen today is the pressure system, which is simple enough. You stop milking her. At the same time, restrict her water supply, feed her dry hay and no grain at all. Pressure builds up in the udder and she will be quite uncomfortable, bleating and making a general nuisance of herself. Be kind and affectionate, but don't give in and milk her. Milk secretion will cease and reabsorption begin. If her discomfort seems excessive, check with your vet. She may have to be partially milked out. In any case, at the end of a week she should be partially milked out to check the udder for signs of mastitis.

The reabsorption cycle is very important, for during the lactation period the doe has been losing more calcium and other minerals through her milk than she has been taking in. During her dry period the doe absorbs enough minerals to ensure her health and that of her kids, and the quality of yet another year's milk supply. She should have a rich diet supplement of minerals. A balanced mix for this purpose is available through your feed dealer.

DEHORNING AND MAINTENANCE

Dehorning. Many goats are born hornless. Those that are not should be dehorned to prevent injury to the goat farmer and to other goats. Horns serve no purpose for the domestic goat, and their early removal is little more than the equivalent of cutting your nails, except that normally it has to be done only once.

The earlier you dehorn the better. To the young, dehorning is no more painful than a slight scratch or burn (depending on the method) would be to you. The same cannot be said for dehorning a mature goat.

Usually you can tell almost at birth if a kid will develop horns. The hair will twist and tuft at either side of the head where the horns will develop. If the hair lies perfectly smooth, the kid will be hornless.

Should there be telltale curls on your kid, wait three or four days, then carefully clip the curls and surrounding areas with blunt-tipped scissors till you have an inch of bare circle. The next day feel the spots where impending horns are suspected. If the skin is loose, it was a false alarm; the goat will be hornless, with at most perhaps a couple of small harmless bumps for ersatz horns. If, on the other hand, the skin is taut and a hard spot about a quarter of an inch or so in diameter appears where the incipient horn will emerge, prepare to act at once.

Dehorning can be accomplished by three means: chemically, through cautery, or surgically. For the beginner, the chemical means is recommended. The traditional method is to stanchion the goat and apply caustic soda, available commercially in a dehorning pencil. First clip the surrounding area closely with scissors and then clippers; grease all the skin approaching the horn bud well to prevent the caustic from burning the skin. Then paint the horn buds carefully with the caustic stick. *Always use extreme caution when handling caustic.* Keep the kid stanchioned for an hour. Before releasing, bandage the head with two oversized Band-Aids, and keep it bandaged for a day to prevent the kid from rubbing the caustic onto other parts of the skin. Don't let the kid out in the rain the first week; water will make the caustic run. That's it, you've dehorned your first goat.

If the horn itself has broken through the skin, surgical removal is advisable. However, neither cauterization nor surgery is a do-it-yourself proposition unless you have helped someone else with it several times first. Get someone with more experience to teach you.

Clipping. Keeping your goat's hair trimmed is an essential part of her maintenance, both for general cleanliness and to discourage lice, ticks, and other parasites. Clipped goats are more comfortable in the summertime too, and a happy goat gives better milk.

The best time to give your goats an all-over haircut is when spring is far enough along so you're pretty sure there won't be any leftover winter nights that might foster colds. But throughout the year, once a month the udder and tail areas should be clipped. This prevents milk contamination and reduces lice, which often initiate their attack by the tail. If you have a severe louse problem, while you're at it rub a little louse powder up and down the spine. However, by brushing the goat's coat carefully two or three times a week you should be able to keep lice infestation down to a minimum, and your goat will love you for it.

Clipping itself is simple enough. The operation and tool involved are similar to those found in the old-fashioned barbershop. (If you have more than two goats, you'll probably find yourself wanting an electric clipper.)

1. Stanchion the goat on a milking stand—but not in the milking shed. Outside where the light is good is best.

2. Pet the goat with your free hand. Comforting her with a few kind words certainly doesn't hurt. Start clipping gently with the clippers. The ones specifically designed for animals will leave just the right amount of stubble.

3. Always clip against the lay of the hair.

4. Make certain to stretch the skin tight when you trim around the udder. Holding your shoulder against the doe will steady her and prevent accidents.

Hoof Trimming. When living in the wild, goats trim their hoofs naturally by wearing them down on hard ground and rocky cliffs. Frolicking in a nice pasture and barn living, on the other hand, don't give them enough wear. So you'll have to do some paring.

Pick a day when the weather is damp, and you'll find the hoofs will be more pliable. Should the hoofs be particularly brittle, apply hoof oil two days before trimming. Your tools are a sharp pruning knife and a medium-grain file.

Stand beside the goat, facing her rear. As with clipping, a steady conversational patter and an occasional petting are good. You're in the country now, and no one's going to look at you strangely if you talk to an animal—farmers know.

Pick up a leg and anchor it between your knees. The center of the cloven hoof has a soft hump, known as the frog. This should not need trimming under most circumstances. If it is overgrown, be extra careful in tackling this area, since it will be very painful to the animal (and probably yourself) if you cut too deeply. Always trim a very thin slice at a time. You can always cut more off—but you can't glue it back on.

The horny edges, or hoof, of the foot should be trimmed so that they are even with the frog. This is why if you trim the hoof regularly you won't have to worry about the frog. It is softer and will wear itself down to the proper level. Check carefully when you're through trimming to see that the foot sits squarely on the ground. If it is uneven, making contact only on the right or left, permanent leg injury may result.

UDDER PROBLEMS

The health and well-being of the goat's udder is, of course, paramount in dairy production. Problems do arise occasionally, but most of them are dealt with by simple prevention.

TRIMMING GOAT'S HOOFS

HOLD GOAT'S LEG FIRMLY BETWEEN YOUR KNEES AS SHOWN. TALKING
TO HER WILL HELP KEEP HER QUIET WHILE YOU WORK.

Dehorning keeps goats from injuring each other. Roomy stables
prevent teats from being trampled. Briars should not be permitted to
grow in pasturage, since they easily cut the udder, sometimes with ensuing infection. And, as opposed to bees and chickens, bees and goats
are not a good combination. The bees seem to prefer stinging the goats
on the exposed udder over all other places, so have your hives at the opposite end of the yard with their entrances facing away from the goat
pasture.

Mastitis. The worst udder problem is mastitis. Actually, mastitis
is a general term for any inflammation of the udder, and not any specific
disease. The reason a multitude of sins fall under the same heading is
that they all manifest themselves in a similar manner—flaky or clotted
milk and a progressively swollen and sensitive udder.

Needless to say, mastitis caused by mechanical injury is not infectious. Bacterially-caused mastitis, usually produced by a *Streptococcus*, is. Avoid giving milk from a doe suffering from mastitis to a kid. This is important to remember, for udder infections tend to be more frequent during kidding.

Mastitis is a problem for your veterinarian to handle, but its early diagnosis is up to you. For this purpose use a strip cup. This is a small container covered with an exceedingly fine mesh. Draw a few drops of milk onto it with each morning's milking. Small flakes or lumps in the milk that become visible on the screen should lead you to suspect trouble. Mastitis causes actual physical changes in the udder and thus can permanently dry up the milk supply. The earlier the treatment—usually antibiotics—the better.

Chapped Udders. Chapped udders, particularly at the teats, are caused by the same irritations that cause chapped hands. Cleanliness and dry milking, that is, not milking with wet hands, should prevent the problem. However, if it does arise, rub glycerin onto the affected areas twice a day and particularly after milking. Dr. Naylor's Udder Balm—worth keeping around just for its folksy nineteenth-century tin box—is even better.

No Milk. If a doe fails to give milk after freshening and the udder swells properly, indicating an ample supply, the diagnosis will probably be an obstruction in the teat—this may also be the cause of "hard," or difficult, milking in later stages of production—or *atresia*, a blind teat which has no perforation at all. The obstruction can usually be cleared without much trouble. Opening a teat manifesting *atresia* is also no major operation. But both should be done by your veterinarian. On the other hand, when the veterinarian comes, never let him go at it alone. Always watch. The more you observe, the more you learn.

Leaking Teats. If you see your doe's teats leaking when she comes into the barn to be milked, milk her more often. Not only do leaking teats waste milk, they draw flies and in general decrease the sanitary conditions of your operation. The kind of milking problem that leads to leakage can also lead to self-sucking. The pressure of the milk becomes painful, and the doe nurses herself to relieve it. This can become a hard-to-break habit that severely curtails your milk supply. Again, the best cure is an ounce of prevention. Milk on a three-times-a-day schedule rather than two if necessary.

OTHER PROBLEMS

Goat Pox. You've heard of smallpox and chicken pox. Well, if you're going to keep goats, look out for goat pox. Not a serious disease, it manifests itself in the same body eruptions as the other poxes. But do not use milk from goats with the pox, which lasts around three weeks, because it's highly infectious. Better yet, have your goats vaccinated.

Parasites. Parasites, external and internal, seem to have a special affinity for goats. Here, as always, prevention is the key word in care.

Many goat parasites make their home in the intestinal tract; they or their eggs are expelled in the manure and transmitted from goat to goat. A clean barn and rotating pasturage minimize the spread of these parasites.

The presence of internal parasites should be suspected when either a goat's stools are looser than usual or constipation sets in and cannot be remedied by the addition of more bulk to the feed. The animal becomes listless and loses weight, all without any accompanying fever. A microscopic examination of sample droppings by your veterinarian or state Agricultural Extension Service should ascertain if stomach or tapeworm parasite infection is the cause. Remember, goats allowed to roam on a large, clean pasture are less likely to pick up parasites than a crowded herd.

Head grubs and follicle mites are two external parasites that afflict goats occasionally. Be on the lookout for unusual skin lesions, scabs, or ruptures, particularly around the eyes, nostrils, and other mucous membrane areas. Your veterinarian should be able to specify their cause and remedy. Once you've observed and treated the different varieties, you'll be able to handle the problem yourself.

All this may make it seem as if you'd better have a direct wire to the vet. Not so. You might well go without seeing him for a year or two. It's just that by being aware of *potential* problems and taking the simple precautions needed to ward them off, you're bound to have healthier goats. And you'll probably be surprised to find that a goat on the farm is really no more complicated to care for than a dog in the city—except that you'll be milking it two or three times a day instead of walking it.

Chickens

You can't hatch chickens from fried eggs.
—PENNSYLVANIA DUTCH PROVERB

About the only thing chickens won't give you is milk. Other than that, they're as close to the perfect barnyard animal as you can find for the commune and for the small-scale farmer. Not only are the chickens themselves and their eggs high in nutritive value, but chickens produce excellent fertilizer, the feathers—mix them with duck or goose down—can be used for stuffing anything from pillows to supersized comforters, and the birds themselves integrate happily with such other country critters as bees and pigs. Besides, they're easy to raise. With the proper setup they can even be left alone on weekends without much worry.

The first thing to consider in poultry raising is the chicken itself. Do you want just an egg producer, or stock mainly for broiling, or a bit of both? Or how about even some show birds? Raising chickens for their plumage was once quite a sport, and varieties such as Mottled Houdans, Blue Andalusians, or the precolored-Easter-egg-laying Araucana, which turns out eggs from green to pink, rival at least female peacocks when it comes to beauty.

SELECTING THE RIGHT BREED

Egg Producers. The Single Comb White Leghorn has been bred over the years to be a top-notch layer. As such it is one of the most popular white-egg producers around. On the average, a Leghorn weighs around four pounds when entering its production stage. This makes it too small to be a family-sized eating chicken. (On the other hand, while the hens are normally considered cooking birds, the young roosters are finger-licking good fried.) The hen's smallness helps make it an efficient producer, since it will eat less and lay as much. A six-pound, 250-egg-a-year bird puts away about thirty-five pounds more food a year than a three-pound, 250-egg one.

Another popular egg-layer is the Minorca, which comes in three varieties: White, Black, and Buff. The Minorca lays a large chalk-white egg that surpasses the Leghorn's both in size and purity of color. It is not as productive a layer, however, and unless you're going to raise a large enough flock for commercial production of extra-large eggs, it would probably be better to stick with the Leghorn.

Meat Producers. Here's where size counts. The Jersey Giant when full-grown will be a nine- to fourteen-pound bird; the Light Brahma matures in the eight- to twelve-pound range. An elegant bird it's not; it's generally clumsy, tripping over everything in the barnyard. The bird doesn't really fly, but it thinks it does.

Both Jersey Giants and Light Brahmas lay large brown eggs—nowhere near as profusely as the Leghorn, however. And because of their size, they are not raised for broiling or frying. It takes them too long to flesh out, making them a bit tougher than modern sensitive tastes call for. As roasters, however, they are unexcelled, and of course there's always the old stew pot.

General Purpose Breeds. Plymouth Rocks, Rhode Island Reds, and New Hampshires head the popularity poll in the all-around class. They produce good-quality brown eggs in quantity, and plump fryers or broilers as well as roasters.

The Plymouth Rock is one of the oldest general farm breeds. The most popular varieties are the Barred and the White (the white is preferred for meat production, since it has no black pinfeathers showing when the bird is dressed). Usually a mature Plymouth Rock weighs in at around eight pounds.

Rhode Island Reds lay big brown eggs of uniform color. Although the eggs are usually larger than those of Plymouth Rocks, the bird itself is around a pound or so lighter.

My own personal favorite and recommendation for the apprentice farmer is the New Hampshire, a graduate of Professor H. A. Richardson's breeding efforts at the University of New Hampshire. Characteristically, it is an early-maturing bird that feathers out rapidly. It lays large brown eggs and in general is quite disease-resistant. An excellent bird for flocks of from twenty to two hundred birds, easily cared for by the beginning farmer.

Show Birds. These are simply for the pleasure of breeding more beautiful fowl, and include, besides those already mentioned, Buttercups, the White or Buff Crested Polish, and still others. Wait until you're really into chickens before tackling these. Once you've raised a few flocks, however, there's nothing like a good show bird to make the old color tube pallor.

STARTING THE FLOCK

There are three ways to start a flock: by buying day-old chicks, started chicks, or ready-to-lay pullets. For reasons of economy, only day-old chicks are of interest to the beginner. To the city dweller, used to supermarket prices, the cost of day-old chicks will seem like the breeders are giving them away.

Buying started chicks would save you the trouble of tending them during the brooding stage; it would also eliminate the cost of a brooder. On the other hand, it greatly increases your chances of introducing disease into the house through infected chicks. This disease potential eliminates the choice.

Ready-to-lay pullets put you into the egg business within a week or less. But since you'll probably be considered a city hick, the chances are you'll get stuck with some poor layers, plus you'll have to worry about vaccinations. And your capital investment will be high. Chicken breeders aren't horse traders, but they come a close second.

Day-old chicks are usually available year-round. Normally they are grouped as Early Hatched (January, February, March), Spring

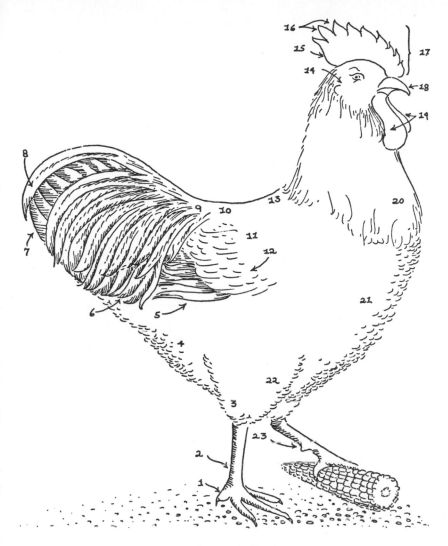

CHICKEN

1. CLAW	8. SICKLES	16. POINTS
2. SHANK	9. SADDLE	17. COMB
3. HOCK	10. BACK	18. BEAK
4. FLUFF	11. WING-BOW	19. WATTLES
5. PRIMARIES (FLIGHT FEATHERS)	12. WING-BAR SECONDARIES	20. HACKLE
6. SADDLE FEATHERS	13. CAPE	21. BREAST
7. MAIN TAIL	14. EAR	22. THIGH
	15. BLADE	23. SPUR

KNOW YOUR CHICKENS

Broiler:	Eight- to fifteen-week-old cockerel
Cock:	Male over a year old
Cockerel:	Male less than one year old
Day-old chick:	Chick shipped to you before its first feeding
Hen:	Female over a year old
Pullet:	Female chicken under one year old
Ready-to-lay pullet:	Female five or six months old, ready to begin her laying cycle
Roaster:	Male up to nine months old
Sexed chick:	Day-old chick that has been classified by sex
Started chick:	Three- to four-week-old unsexed chick
Started pullet:	Eight- to twelve-week-old female

Hatched (March, April, May), Late Hatched (June, July, August), and Fall Hatched (September, October, November). For your first venture into chicken-raising, spring hatched day-old chicks are recommended, since given the good weather conditions that time of year, they are easier to raise. They will reach the egg-production stage in five to six months and you'll have a steady supply before winter sets in.

Pullorum-Clean Chicks. Pullorum is a highly contagious disease caused by *Salmonella pullorum* bacteria. It is fatal to chicks under three weeks of age. There is no cure, and half the time there are not even any symptoms until the final result: a mass of dead chicks. White diarrhea, when present, may be a telltale clue, but manifestation of this symptom is no real help, because once infection has set in, there is nothing that can be done about it. Since the disease is transmitted from infected hens to their eggs, and the eggs hatch as infected chicks, effective control lies with the adult bird.

Always make certain that your brood comes from "pullorum-clean" or "pullorum-passed" stock. You might think you left the tricky word games of Madison Avenue promoters behind when you came to the country, but watch out for breeders wanting to sell "pullorum-tested" stock. The phrase doesn't mean a thing except that the parent stock was tested—and in all probability found to have the disease. The word "clean" or "passed" must appear on the statement of warranty somewhere, even if it's in the more lengthy combination of "pullorum-tested clean."

Usually you will find it most convenient to start with twenty-five chicks, since this is the minimum number of pullets most dealers will ship. Chances are there'll be early mortality of around 20 percent, and 5 to 10 percent will turn out to be cockerels, due to sexing errors, so you will end up with a laying flock of fifteen to twenty hens. This will give you about a dozen eggs a day.

You can also start with straight-run chicks; that is, chicks which have not been sexed by the dealer. They are cheaper; however, with twenty-five of them you're likely to end up with only ten layers. Half will turn out to be cockerels and four or five will die.

At the top of the small farmer's scale are flocks of 250 to 300 birds. A flock of this size is not at all unfeasible, even on a part-time basis, though you'll probably want to wait a couple of years before tackling one this big. It will integrate well into your farm plan, producing a good quantity of fertilizer for your garden (about 125 pounds of usable manure a year per bird) and a large enough quantity of surplus eggs and meat chickens to justify finding a market for them, earning you a small cash surplus. All this without taking so much time as to dominate your daily schedule. Much above three hundred birds, however, and you're involved in a full-time operation. A commercial flock doesn't become really profitable unless you have layers in the thousands or even tens of thousands. Not only does it require considerable investment, it makes you a chicken factory manager, and in many ways takes away much of the attractiveness of general farm life.

THE HEN HOUSE

Chickens used to live houseless in the wild, but that was almost as long ago as man did. The modern chicken needs protection not only against winter cold, but summer heat as well. A chicken can't sweat, having no glands for this purpose, and so is very prone to heat exhaustion. Hence the need for something more than a nest to call home.

Site. Select a good location for your chicken house. Sticking it behind the barn or just any old place can cause a lot of difficulty later. Proper drainage and air circulation are important. Therefore, if a knoll or small hill is available, build there. But not at the top of the knoll; the house would be too exposed to the elements. And not at the bottom either, because there the house would be difficult to ventilate with natural air flows. Cold, moist air settles in depressions and it would in all likelihood be quite damp. The best spot is halfway down the knoll, on the south side where the sun shines more. If you have no choice because your land is flat, make at least a small artificial knoll of gravel if possible.

It's also a good idea to build the house beneath the shade of a tree or two. The trees should be quite large, since low-hanging branches will foster dampness. For the same reason, the house should be located at least eight feet from the trunk.

A human factor to consider when locating your chicken house is the nearness to your own abode. Although you probably won't want the chickens right under your bedroom window, neither will you want them on the back forty where the long daily trudges to the coop might tempt you to take undesirable shortcuts in their care and feeding.

House Rules. Of primary importance in a chicken house is cleanliness. Chickens are not naturally clean, and when raising them within the confines of a house, this can lead rapidly to disaster. The best preventative against chicken disease is careful and regular cleaning of their environment. All equipment such as watering fountains and feeding troughs should be free-standing, as opposed to bolted down or built in, so they can be easily taken out and hosed down. If you're building your own hen house, it should be constructed to facilitate easy cleaning and disinfecting. Avoid building feeders into the corners or leaving many raw, unfinished edges. The house should also be so constructed or located that the chickens can range on different parts of the yard each year. A simple way of achieving this is to build a front and a rear chicken entrance to the house and have alternate yards. The year one yard is off-limits to the chickens, use it and its rich accumulation of manure as a garden. Another way is to build the chicken house on skids, so it can be moved from one location to the other.

Ventilation is another key factor in a healthy hennery. If you're building a house from scratch, be sure to include enough windows for plenty of ventilation. On the other hand, don't be discouraged from using a small outbuilding already on your property simply because it has no windows. Put them in. It's easier to install windows than to build an entire house. To be effective against the collection of moist air and the natural poisonous gases given off by the birds themselves the house should have, besides its chicken entrances, windows that can be opened on three sides, with the door for you on the fourth. All windows must be covered with quarter-inch wire-mesh screening to keep sparrows out. Sparrows love a well-stocked, ready-made birdhouse and are no respecters of species.

Humans need vitamin D regularly and so do chickens. Hence sunlight is very important to them. It is also an aid in cleanliness, being an excellent germicide, and in general gives comfort to the birds. If possible, the largest windows should therefore face south or southeast to maximize the sunlight.

The size of your hen house will vary, of course, with the number of chickens you keep, but it's a good idea to have extra room available for future use. You may find that raising chickens suits you. Although you probably won't want to go into it on a commercial scale, after a year or so you'll probably find that selling surplus eggs, either from a roadside stand or directly to a store or hotel, yields ready pocket money. Also a straight barter agreement of milk for eggs with a local dairyman is handy if you don't want to get into the milking routine required by goats. Whatever the reason, your flock is bound to grow with each ensuing year, so plan ahead.

The standard rule-of-thumb in hen-house zoning is four square feet of floor space per layer. With a ten-by-ten-foot house you have room for twenty-five layers; a house ten by twenty feet expands your capacity to fifty; and so on.

FURNISHINGS

The basic equipment needed for a chicken house is simple enough. Food hoppers and water fountains for intake, nests and dropping pits for output, and roosts for sleeping take care of the lot.

Food hoppers are usually designed with either a flat or a V bottom. The latter is more convenient and easier to construct. Just take two boards and nail them together lengthwise at a ninety-degree angle. Across the end of this extended V nail two square leg pieces big enough to keep the trough upright and slightly off the floor and the feed from spilling out over the ends. Allow about one foot of hopper length for every two birds, with the hopper accessible from both sides. It's better to have two or three short hoppers than one long one. They're easier to handle for cleaning purposes, and also permit the flock to scatter rather than all having to crowd to one trough to eat. Separate hoppers should be provided for oyster shell and for grit. The hoppers are usually placed along the front of the house, away from the roosts, parallel to the wall.

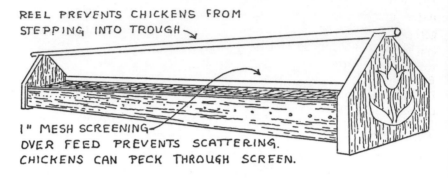

REEL PREVENTS CHICKENS FROM
STEPPING INTO TROUGH

1" MESH SCREENING
OVER FEED PREVENTS SCATTERING.
CHICKENS CAN PECK THROUGH SCREEN.

CHICKEN MASH HOPPER

Water fountains must be designed to prevent the birds from tipping them over. A plain pail set at an angle in a stand will suffice for small flocks. However, the commercial water fountains available require only a minimal investment. They are well worthwhile from the dual standpoints of sanitation and convenience.

Roosts must be provided. A good arrangement is two-by-four-inch planks placed on edge, rising in bleacher fashion—two feet from the floor for the first one, two and one half feet from the floor for the second, and so on. The distance between roosts should be one foot. Allowing one foot of roost per bird will avoid crowding. Crowding fosters disease, poor feathering, uneven growth within the flock, and cannibalism or picking, all undesirable characteristics.

The Nest. If you don't provide them with nesting space, your birds will lay their eggs in any corner they find convenient. Not only does this make extra hunting for you, but chances are you won't get any of the eggs. The hen-house floor will begin to resemble a trampled uncooked omelet. Even those few eggs you rescue will be of questionable age, and may have been reheated several times over by any number of motherly-minded hens who found and sat on them before you collected them. Cracking a month-old oft sat upon egg into the morning frying pan is not the most pleasant wake-me-up.

Although you can construct simply an open-sided nest, the nesting box, in which all sides except the front are enclosed, is much preferred.

The nesting box furnishes a dark, homey compartment which minimizes egg-eating, breakage, and soiling. Row housing is nicely adapted here. If you make the boxes long enough to permit several compartments for individual nests, Pullman fashion, a number of chickens can nest to-gether. They seem to enjoy a good, gossipy coffee klatsch, and a happy hen lays more eggs.

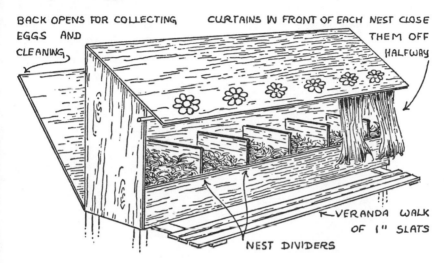

ROW NESTS

CHICKENS WILL LAY BETTER WHILE TALKING TO EACH OTHER.

Nesting boxes two feet wide and one and a half feet high, made of plywood, are cozy. Build them in five-foot or ten-foot lengths, subdividing the box with eight-inch-high dividers every foot to provide individual nests. The bottom of the box should be three feet off the floor, so supply a walkway in front where the hens enter. This permits them to strut back and forth in choosing a suitable unoccupied nest. Parallel one-by-one-inch wood strips, with a one-inch space between them, running along about a foot below the nest openings, make an excellent balcony for this purpose. A clean nest equals clean eggs, so change the straw or hay every two weeks. For real efficiency, many farmers design their nest boxes so they can be opened from the back. This makes them more convenient for cleaning.

Housekeeping. With chicken-raising, most of the changes have occurred in the bird itself. Pure breeds and hybrids have been developed to produce more and bigger eggs in layers and heftier, more tender birds in broilers. The only really major breakthroughs in management of the hen house itself have been the concepts of deep litter and dropping pits.

Like many great ideas, deep littering and pits probably evolved as products of laziness disguised under the name of efficiency. The point is, not only do they save time and labor, but they also promote the general welfare of your hens.

Under the old system, dropping boards were placed beneath the roost. These were scraped and cleaned once a day. Nowadays, a pit area is boarded off under the roost and covered with chicken wire. This permits the droppings to fall through, yet keeps the birds away from it, thus minimizing disease. With dropping pits, the daily scrub-down is replaced by a sprinkling of superphosphate on the "night soil," which adds to as well as preserves its fertilizer value by fixing the nitrogen. The pit is cleaned out once a week; during cold weather this can even be reduced to once every ten days.

Your compost heap gets the droppings. Never apply fresh chicken manure directly to your crops. In the case of tubers, like potatoes, don't even use it on land you intend to plant the same year. It will burn the seeds and ruin your crop. Always compost first.

In deep littering, or built-up litter, a four-inch layer of straw, ground corncobs, sawdust, peanut shells, or wood chips is rapidly built up on the hen-house floor, except, of course, in the dropping pit, over a period of three weeks or so. After this initial base is laid down, it is topped weekly with a thin layer of fresh litter and superphosphate. The litter is allowed to accumulate and, like old wine, improves with age. Microorganic development in the litter makes it a rich food source. Deep-litter-raised hens are less prone to disease, since the litter supplies vitamins and antibiotics. Cannibalism, brought on by an imbalanced diet, is noticeably reduced or eliminated. Twice a year you clean the floor and start a new litter cycle. The old litter goes straight to your compost heap.

The poultry house should be whitewashed inside once a year to help keep down disease and parasites. Litter should be checked to make sure it does not get damp. Prevention is the best cure here too, for once the litter becomes wet, it must be replaced. Mash hoppers and water fountains should always be clean, and never be empty. Once a month they should be sterilized.

PREPARING THE BROODER

For baby chicks, the poultry house should be as clean as possible. Be particularly careful if you're going to move them into an old abandoned chicken house that you have renovated without knowing anything about the previous tenants. Scrape the floors. Sweep floors, walls, and even the ceiling. Hose the whole house down with water. Once it has dried, sterilize the whole place with disinfectant. Whitewash the walls and ceiling. Sterilize thoroughly any used hoppers or water fountains you intend to use.

The little chicks may seem rather lost in such a big house, but don't worry, they'll grow with a rapidity you won't believe. Don't worry about the house being cool either. As long as you have good temperature control within the brooder itself, the chicks will suit themselves about the temperature they need, heading for the warmth whenever they feel the world outside is too cold.

A brooder can be made simply from an inverted box, open at one or two sides, with the openings covered by curtains. Or invert an old wash tub, prop it up off the floor, and ring curtains all around the bottom.

The heat supply can be simply an infrared light bulb in a porcelain socket, with a pie-pan reflector. Unfortunately, however, this setup has no temperature control. Also, even new light bulbs sometimes blow. You not only have to constantly check the temperature, but you risk losing the incipient flock to a faulty light bulb.

A commercial heating unit is well worth the investment. It is not that expensive, and will last many years. There's no need to buy a whole brooder, which represents more of a capital outlay, although it would save work.

The young chicks will tend to cluster together under the brooder; even so plan on ten square inches of heated floor space per chick. Start the heater two days before the chicks arrive to test it for steady and accurate temperatures.

Cover the floor litter with heavy paper. Leave it there for a few days after the chicks arrive to help them learn the difference between litter and food.

The brooder itself should be surrounded by a twelve- to fifteen-inch-high wall—round, not square—of linoleum, tin sheeting, or cardboard heavy enough to withstand the pummeling of chicks. This encirclement, or chick guard, should be about two feet away from the

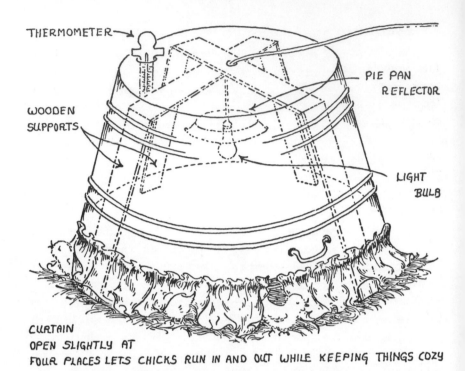

THERMOMETER

PIE PAN
REFLECTOR

WOODEN
SUPPORTS

LIGHT
BULB

CURTAIN
OPEN SLIGHTLY AT
FOUR PLACES LETS CHICKS RUN IN AND OUT WHILE KEEPING THINGS COZY

CHICKEN BROODER MADE FROM GALVANIZED WASHTUB

edge of the brooder for the first day. Gradually move it farther and farther away. After the second week it can be removed entirely. The guard prevents the chicks from straying too far from the brooder and, because it is round, there is no danger of a group of chicks trapping themselves in a corner and smothering each other.

Feeding during the first week is best done from paper plates. After that you'll want to switch to small troughs; for young chicks these should be sterilized every day. Make sure the water fountains, also to be sterilized daily, are nontippable. Have two one-quart fountains ready for the first two weeks, then switch to the gallon size. It's best to set the fountains up on low, wire-screened platforms to lessen water fouling by droppings and litter. Plan on locating the water fountains halfway between the brooder and the feed. This will ensure the likelihood of the chicks drinking before and after eating.

FEEDING

A balanced chicken diet should contain proteins, carbohydrates, fats, minerals and vitamins, and, of course, water. Water is not only needed for blood, which acts as a carrier for the other nutrients to different parts of the body; it regulates the body temperature as well. A chicken's body itself is over 50 percent water, an egg almost 65 percent.

The basic constituents of poultry feed are: whole small grains, such as wheat and rye plus cracked corn, called scratch feed; and finely ground grain and grain by-products with protein and mineral supplements, called mash. Yellow corn is usually the chief grain fed poultry, at times composing as much as 50 percent of their diet. Yellow is chosen over white corn because of its high vitamin A content, absent in the white. Wheat and oats are valuable additions to a corn diet. Some chicken farmers use barley when it's cheap; however, it must be "debearded," that is to say, have the awn removed, or the chickens will choke to death on it. Be sure you can recognize a shaved barley grain before you feed it to your flock.

FEED FORMULAS

Chick and Growing Mash
   20 lbs. yellow corn meal
   20 lbs. wheat bran
   20 lbs. pulverized oats
   10 lbs. fish meal or 15 lbs. meat scrap
   10 lbs. alfalfa meal
    2 lbs. oyster shell meal
    1 lb. cod liver oil
    1 lb. salt

Chick Scratch
   100 lbs. finely cracked yellow corn
   100 lbs. cracked wheat
    50 lbs. pinhead oats

Growing Scratch
   100 lbs. medium cracked yellow corn
    50 lbs. whole wheat

Laying Mash
> 100 lbs. ground yellow corn
> 100 lbs. ground heavy oats
> 100 lbs. ground wheat
> 75 lbs. fish meal
> 25 lbs. dried milk
> 15 lbs. ground oyster shells
> 4 lbs. salt

Laying Scratch
> 100 lbs. coarse cracked yellow corn
> 60 lbs. whole wheat
> 20 lbs. whole buckwheat

A protein supplement of fish meal is probably qualitatively the most dependable, containing on the average 60 percent protein. Meat scraps, milk, dried or fresh, and oil-cake meals from soybeans, peanuts, and the like also offer a good source of protein.

Salt aids digestion and stimulates appetite. It is usually added to the mash at the rate of 1 percent of the whole. Edible steamed bone meal supplies phosphorus; and oyster shells, limestone, or calcite supply calcium to the diet. Without these the number of eggs would be greatly reduced and their shells so thin as to break at the slightest excuse.

Grit, composed of sand or small pebbles, is essential to the chicken's diet. You might not like sandy spinach, but hens thrive on it. A chicken doesn't chew, and its digestive system is unable to break down most of the food it eats unless there is plenty of grit to grind the food particles down to size.

In the summer feeding your hens greens is no problem. Almost any scraps from your vegetable garden will do admirably. Carrot tops, kale, and other leafy wastes are excellent. Swiss chard is good but, along with rape, it should be fed sparingly to laying flocks, since both of these cause egg yolks to turn very dark. During the winter cabbage is one of your best greens. The birds like it and it's easy to store. Sprouted oats also make an excellent winter green. Farmers were using oat sprouts long before Chinese cooking introduced the bean sprout to the American diner. Sprout the oats the way you would beans, except on a larger scale. Place the oats in a bucket filled halfway with water as hot as your.

hand can take it. Let soak overnight, then drain and spread them out in a one-half- to one-inch layer on trays. Sprinkle with warm water whenever they seem dry. Moisture and warmth are the keys to successful sprouting. Harvest the shoots when they are about one-half to one inch long, usually on the fourth or fifth day. Feed in the middle of the day, but no more than the birds will clean up in fifteen to twenty minutes.

ARRIVAL OF THE FLOCK

A cheeping box arrives filled with down balls. Rush them to the ready and preheated brooder and release them beneath the cover, known as the hover. Don't forget to count them as you unpack. Make sure there's feed and water for them. The chicken guard should be in place.

Now you might think baby chicks will eat and drink when they're hungry. They won't, until you play mother hen and teach them. Tap your finger on a feed plate—scatter oatmeal over the feed, it has eye appeal for chicks—to attract their attention. They'll soon begin their own pecking. They learn fast, but not instantly, so be prepared to repeat the process the first two days. Pick up the chicks individually and dip their beaks in the water a few times, letting them shake it off their faces in between.

BROODER AND DEVELOPMENT SCHEDULE

First Week
Temperature: 90–95°F. in brooder
Mash: Constant supply sprinkled lightly with sand on paper plates
Scratch: Fine cracked corn fed twice daily starting second full day
Water: Warmish to room temperature, 2 1-quart fountains per 25 chicks, with a drop of concentrated potassium permanganate solution added to each quart of water, to keep them healthy

Second Week
Temperature: 85–90°F. in brooder
Mash: Constant supply sprinkled with sand in low trough
Scratch: Fine cracked corn fed twice daily, heavily in the evening
Water: Room temperature, continuous supply in fountains, with 1 drop potassium permanganate per quart
Greens: Fresh, fine-cut, all that the chicks will eat in 20 minutes

Third Week

Temperature: 80–85°F. in brooder

Mash: Constant supply with sand in low trough

Scratch: Switch gradually over the week from fine cracked corn to medium cracked corn, fed twice daily, heavily in the evening

Water: Room temperature, and make sure to add extra fountain space as the birds increase in size

Greens: Fresh, fine-cut, all they'll eat in ½ hour

Exercise: Let chicks outdoors on clean ground daily after the dew has dried off the grass and as long as there is no rain

Fourth through Seventh Weeks

Temperature: 75–80°F. in brooder

Mash: Constant supply with sand in low trough

Scratch: Medium cracked corn fed twice daily, heavily in the evening

Water: Cool, continuous supply

Greens: Fresh-cut, all they'll eat in ½ hour

Exercise: Let chicks out daily on clean ground in dry weather; start training to roost by laying roosts on floor, after a few days raising them slightly, and so on

Eighth Week

Remove brooder heat

Add oyster shells and grit instead of the finer sand to diet, in separate troughs

Get birds to eat more scratch and less mash

Use regular coarse cracked corn for scratch

Let pullets out at daybreak to minimize picking

Vaccinate pullets for fowl pox if necessary in your region

The brooder should be taken out by the end of the eighth week, although this deadline for removal can be extended to the tenth week in northern climates if there's a chance of severe late cold. By this time, during the day all windows and any ventilators should be kept open. After the tenth week, or once the chickens have feathered out fully, the windows should be kept open both day and night.

Make sure the litter stays dry and loose. If wet or caked, replace it.

Range your birds. The best range is a poultry yard on which no chicks have been raised for two years. This is why you want your chicken house to have two or more exits. The first year is no problem, of course. But the second year you'll want to range them from the opposite side of the house. Using the old range for a garden is highly advantageous, since it has been well fertilized the previous year.

Shade is important on the range. And the best shade is natural. Jerry-built tin-roofed shade areas are never as cool as those shaded by trees and bushes. Growing plants, through evaporation, provide the most comfortable escape from direct sun. A pasture of sunflowers and/ or corn is excellent, but the plants will have to be four or five feet high before the chickens are let loose or you will shortly have no plants left.

EGGS AND THE LAYER

Between the ages of five and seven months your hens should begin laying. Their house needs an extra lively cleaning at this time, followed by the installation of the nest boxes, lined with straw. Change the straw every two weeks for the sake of continued cleanliness. It might be a good idea to buy a few glass eggs to put in the nests. The birds will take a hint.

You don't need a rooster to get eggs. In fact, if you're going to want to store the eggs for long periods of time, say, two to three months, they

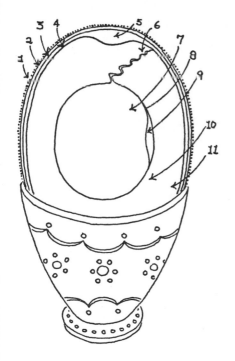

INSIDE THE EGG

1. BLOOM
2. SHELL
3. OUTER MEMBRANE
4. INNER MEMBRANE
5. AIR CELL
6. CHALAZA
7. YOLK
8. YOLK MEMBRANE
9. GERM SPOT
10. THICK WHITE
11. THIN WHITE

should not be fertilized or they'll spoil. Any roosters from your batch of day-old chicks can be put in the stew pot early in the game unless you expect to breed chickens in the future or want a country alarm clock.

A flock of laying hens is best fed by the continuous supply method. For this, have a trough of scratch and one of mash always available. Check daily to see that the troughs are filled enough, and stir up the mash twice a day if you get the chance. Allow a foot of trough, accessible from both sides, for every two birds. Water, of course, must also be constantly available. Continue the daily feeding of greens. Then too, any table scraps you have to offer make an interesting addition to a layer's diet. Set up a separate trough, divided into two sections, containing oyster shells and grit for the hens to peck away at when they choose.

Commercial egg factories confine layers to the house permanently. However, for a small-scale operation it's best to let them range all afternoon. Chickens are natural peckers, and they will compensate for any accidental diet imbalance by scratching up the appropriate food from the ground wherever possible. But don't let the chickens out of the house till noon. They lay their eggs in the morning, and the noon recess hour will keep you from having to go through an Easter egg hunt every day.

Gather eggs twice a day, early in the morning and just after you let the hens out, during laying season. Use a basket, preferably the open wire type, which lets the eggs cool down rapidly. Eggs can be stored in a cool, moist basement or root cellar, but not near onions or other odoriferous objects. A refrigerator is fine if it has the space. (By the way, an ultra-fresh egg won't truly set. It has to be at least three days old before you can hard-boil it—and expect it to be hard.)

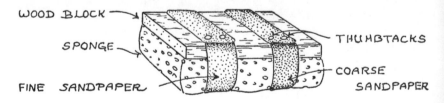

EGG CLEANER

WIPE EGGS OFF AFTER COLLECTING. USE THE SANDPAPER LIGHTLY ON HARD-TO-GET-RID-OF DIRT.

In the thirties hens producing fifty to a hundred eggs a year were considered pretty good layers. Today a chicken not laying 250 to 275 eggs a year is laughed right out of the hen house and into the stew pot. Well-kept Leghorns or Rhode Island Reds should lay around three hundred a year. Even a mixed meat-and-egg producer like the New Hampshire lays 250-plus if well managed.

LAYER/NON-LAYER CHARACTERISTICS

Part	High Producer	Low Producer
Comb	Red, full and velvety	Shriveled and pale
Eyes	Bright and prominent	Dull and sunken
Skin	Thin, no hard fat	Thick, with underlayer of fat
Back	Broad and long	Narrow
Body	Full and broad	Shallow
Plumage	Ragged and worn	Well-kept but dull

Egg production from a developed laying flock is at its nadir in January, peaking in April to May before cycling down once more. Keep a production record of the number of eggs laid, pounds of feed and its cost, price of eggs sold, if any, number of chickens eaten, natural mortality, etc. Many feed companies supply free record sheets, or you can design your own. The record serves two purposes. First, it gives you an overall picture of what's happening in your coops, and secondly, it tells you the economics of the venture. You can't expect to make much money on a small flock; probably you're not interested in making money on it. On the other side of the barn, you don't want to be paying twice the price of store-bought eggs either. Careful records will tell you where you're heading.

The Broody Hen. Beware of the broody hen in spring and summer. Birds that want to set, or brood, will stay on the nest day and night, clucking proudly if you approach. This type of maternal behavior is not appreciated in laying flocks, as it cuts down production. Take the protesting hen from the nest at once and confine her to a small, well-lit coop in a high-activity area such as the entrance of the chicken house until she decides that all that sitting around is for the birds and joins the crowd. But don't forget to tend to her needs along with those of the rest of your flock. Confinement will do the hen no harm; deprivation certainly will.

If you have a broody hen around and some duck eggs as well, let her brood on them to her heart's content, as an ersatz mother duck. Or, if you're going to breed chicks, now's your time.

BREEDING

For a breeding flock, one rooster to a harem of ten to twelve hens is ideal. However, note that Leghorns, Andalusians, Minorcas, and other light Mediterranean breeds make lousy mothers. They'll set for a week, and then decide to skip the nest to see what's happening out in the yard, forgetting entirely about their impending brood. So stick with a breed like the New Hampshire if you want to raise your own chicks.

Collect the eggs from your laying flock as usual. When you spot a broody hen, however, give her a nest of her own away from the crowd and preferably in a private brooding house. Slip up to fifteen of that day's freshly collected eggs under her. Remember, both when moving the hen and when slipping her the goodies, that a setting hen, whether there are any eggs under her or not, is sure you're after them, and will peck at you.

A setting hen will get off the eggs only once a day and at most for half an hour. So be sure food and water are right outside the nest so she can eat quickly before dashing back to her lumpy Sealy.

Once the eggs hatch, confine the mother hen to a three-by-three-foot cage with holes at the bottom just large enough for the baby chicks to run in and out. This prevents the hen from wandering too far afield with her young. The caged mother hen works on the same principle as a brooder, except that the heater is hen-powered.

If you have cats, do keep in mind that the all-time favorite cat food is week-old chicken fricassee on the leg. Once a cat has tasted this delicacy, you'll never break it of the habit. Remember the old saying, "Shut the barn door before the cat gets in."

PARASITES AND DISEASES

Poultry raised on freshly ground, organically grown feed and quartered in houses using the built-up litter system are two steps ahead of the average bird when it comes to staying healthy and laying happily. The third step, again a preventative one, is continued good sanitation, a factor that cannot be stressed too often because of the urbanite's visions of smelly old barns with chickens scratching about wherever they please. A basic cleanliness checklist would include the following reminders:

1. Make sure your day-old chicks come from pollorum- and typhoid-clean parent stock.

2. Vaccinate all your pullets against fowl pox and laryngotracheitis where necessary.

3. Raise your day-old chicks apart from your flock of adult birds if you already have one.

4. Rotate the range for your poultry yearly.

5. Make sure your dropping pits are covered with wire screening.

6. Avoid used equipment such as crates and feed bags. Water fountains and troughs should be sterilized once a month for mature birds, once a day for chicks.

7. Burn all dead birds, and offal from slaughtered ones.

8. Never spread poultry manure as fertilizer where your chickens will be allowed to range the same year or, better yet, even the following year.

Even the best-tended flock, however, will have its occasional bout of sickness. Most of the more serious poultry diseases, however, are prevented simply by vaccination.

Fowl Pox. Highly infectious, the pox is most common among chickens in fall and winter. Wartlike bumps on or around the comb, wattles, or eyes are symptomatic. Mortality ranges from zero to complete, depending on the epidemic's severity. Vaccinate chicks at six to eight weeks of age to prevent outbreak. Complete instructions for this simple operation come with the vaccine. Vaccination is not necessary in a region where fowl pox is known to occur only rarely.

Laryngotracheitis. Another highly infectious disease that affects the respiratory system. Coughing, gasping, and repeated shaking of the head in an attempt to clear the trachea are symptomatic. Mortality is high. Vaccination is suggested in the case of a local outbreak, since there is no cure.

Coccidiosis. Caused by a microscopic parasite taking up residence on the wall of the intestinal tract. Heavy mortality in young birds. A droopy look and diarrhea with blood are symptomatic. No cure. In case of outbreak, isolate obviously infected birds. Change litter and sterilize all equipment daily for a week. This should break the reinfection cycle.

Other Intestinal Parasites. Tapeworms, roundworms, and pinworms all are hard to diagnose; only a post-mortem will positively determine the existence of these parasites. External symptoms include paleness (yes, a chicken can look pale; you'll notice the difference as you get to know your fowl), thinness, and a stunting of growth. Egg production may be severely curtailed. No real cure. Proper sanitation, however, should prevent occurrence in most cases.

External Parasites. If you're not familiar with lice and mites, here's where you will get an introduction. In the old hillbilly movies you've seen, those chickens scratching around aren't scratching for nothing. They have lice, or mites, or both.

Lice are one of the most common poultry pests. Not only that, they come in eight popular varieties. But don't worry about yourself, these lice only go for fowl. If your birds look haggard and egg production is down, pick up the worst-looking bird, spread the feathers under the wings and check the skin of the wing pit. If she's got lice, you'll see them crawling. Again, good sanitation is the best preventative. Delousing can only be done with insecticides unless you are going to pick them off one by one. Standard delousing methods include putting pinches of sodium fluoride on the breast, back, underwings, neck, and around the vent. Another method calls for painting 40 percent nicotine sulphate solution on the roost. Nicotine sulphate is *highly poisonous,* so shoo the chickens away and don't close the chicken house windows for at least two hours after use. Personally, I prefer simple good sanitation, and putting up with an occasional louse or two. If you keep the house clean, a few lice shouldn't cut into egg production. The birds will rid themselves of a small population of them if you have a bare dry spot of soil on their range. They'll get right down in the dust and give themselves a good, long, natural dry bath, which works wonders for the problem.

Mites are an even more common poultry pest, and a bit more persistent. To get rid of them, the roosts can be painted with a disinfectant. Carbolineum, a coal-tar product containing carbolic acid, is often used for cleaning roosts since its residual action lasts for a year. However, make certain it has dried thoroughly before letting the chickens back in. Paint in the morning, safe by evening—supposedly. If you don't want to use carbolineum, crankcase oil works well. Dilute 50 percent with kerosene. Paint once, repaint again after two weeks, and the protection should last six months. Since mites have to crawl along the roosts to get from hen to hen, destroying them on the roosts effectively eliminates the problem.

Another way of minimizing the mite and louse problem is to crumble up a handful of dried tobacco leaves in each nest. This usually keeps the parasites under control. Of course, that silly old rooster that never sets is left scratching.

THE LAYER AND THE NONLAYER

Once a year your birds will molt. Old feathers are shed and replaced by new ones. This is no indication of illness, merely a natural change of wardrobe. However, molting time can help you in selecting which birds to keep as layers and which will end up in the roasting pan. A bird that molts late in the year (September or later) is a good layer. A bird that starts molting before September won't finish any sooner than the late starter. And during molting a hen stops laying.

MEAT CHICKEN

Chickens that have completed a year or more of egg-laying before being deselected from your production flock should go into the stew pot because they'll be fairly tough by then. When the cockerels you received in your shipment of day-old chicks are eight to twelve weeks old, they will weigh two to three pounds. At that weight and age they are considered broilers, the most tender of eating chickens. When they reach three or four pounds at the age of twelve to fourteen weeks, they are fryers. At four pounds plus and fourteen weeks to six months, the cockerels are classed as roasters. Anything over that and you have a stewing bird again.

DRESSING

Before killing a chicken, separate it from the flock and give it no food for at least twelve hours. Water is allowed. Withholding food will make cleaning the bird a lot easier by ensuring that the crop—the sack in a bird's gullet which predigests the food—is empty when you extract it.

The modern method of killing a chicken is to cut the veins in the neck and then pierce the brain. This eliminates the reflex movements that can occur for some time after a chicken is properly dead, and which are responsible for the traditional image of a headless chicken galloping around the barnyard. This method, however, requires a fair amount of skill and knowledge of bird anatomy. It is not recommended for the apprentice.

The Chinese method is also not recommended for beginners. Here you tuck the pullet under one arm and, squatting so its neck is over a bowl of cold water, stroke the bird to keep it calm while slowly slitting its throat—a little deeper with each stroke. Well done, and the bird just looks up and smiles as it passes into oblivion. The blood and water are saved for soup; the cleaned bird is served with head and feet.

The old method is still the best—that is, a sharp ax applied blade downward to the neck of the chicken so as to separate the head from the body. Hold both the bird's legs in one hand, take hold of the long feathers of each wing and pull them back to meet the legs. Holding the legs and wings together with one hand will keep the wings from fluttering after decapitation. Place the head on a block and with one swift, forceful chop, as close to the head as possible, lob it off. Hold the carcass over the compost pile, or let it drain into a pan and dispose of the blood the same way. The carcass, once the blood has stopped flowing and all muscle reactions have halted, is ready for plucking.

Chicken feathers can be loosened by either semiscalding or hotscalding. Hot-scalding is quicker and easier, but it may also give the skin a somewhat cooked appearance before you actually cook it—sometimes, if you're careless, even causing pieces of skin to fall off.

For hot-scalding, water close to boiling is used. For semiscalding, the temperature should be kept between 128°F. and 132°F. Since semiscalding does not loosen the feathers as thoroughly, however, they must be pulled out in tufts.

If you intend to eat the chicken the same day it's dressed, when it will have the best flavor, use the hot-scald method. You'll need about three gallons of close to boiling water. Take the carcass by the feet, dip it so all the feathers are fully submerged, and swirl it around for half a minute. Lift it out. By dipping the bird in cold water immediately after the scalding, most of the negative factors of this method will be avoided. Rub off the loosened feathers. Pluck all pin feathers still left, and cut off the feet.

Eviscerate, or draw, the bird. First, slit the skin where the neck joins the body, being careful not to cut too deeply lest you slice open the crop, or first stomach. Pull out the crop, a sac at the end of the esophagus in which food is normally stored. You'll find it located slightly to one side of the esophagus, which comes out with it. Yank out the windpipe.

Cut a small circle around the vent. Then cut down from the bottom of the circle toward the breast as far as the tail bone, making an opening just large enough to insert your fingers and extract the intestines, gizzard, liver, kidneys, heart, and lungs. The last are close to the backbone; be sure you've reached them. Remove the gall bladder from the liver, cutting away a corner of the liver, if need be, rather than rupturing the gall bladder and spilling bile all over the place. Bile tastes vile, and a little goes a long, long way. To clean the gizzard, hold it so you can slice along the edge as if you were opening a clam. Cut about halfway through, pry apart, and spread it open. Peel out the inner membrane and discard the gravel you find, which is there to grind the food so the chicken, which has no stomach acids, can digest it—that same gravel in your mouth would also grind your teeth.

Save and use the heart, liver, and gizzard—the giblets. They're very nourishing. Rinse the inside of the bird under cold water and your chicken is ready to cook.

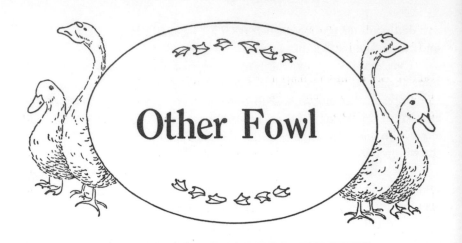

Other Fowl

Man comes and tills the field and lies beneath,
And after many a summer dies the swan.
—ALFRED, LORD TENNYSON

CHICKENS aren't the only useful domestic fowl on the farm. Ducks and geese give you variety in meat and eggs, as well as the highest quality down. And swans, though not a productive addition, personify beauty itself.

DUCKS

Among the hardiest and easiest to raise of all fowl are ducks. For the most part, they are kept for their meat, though there are duck-egg connoisseurs, and the down and feathers make fine pillows. Should you have a pond on your place, the decorative breeds of duck are a pleasant addition. They also help cut down the algae, which does not aid fish growth the way plankton does.

BREEDS

For meat you'll want the Pekin, an all-white duck with a yellow beak. Black or partially black beaks are sometimes seen, but a black-billed duck is considered inferior stock.

While Pekins will lay quite well, if you really want an egg-producer, the Indian Runner is your breed. Coloration ranges from white to white and fawn to pencilled. Since these ducks grow to only half the size of Pekins, the roast duck will be a small one. On the other hand, they not only lay more eggs, but ones of better quality. An Indian Runner will give you almost as many eggs a year as a lazy chicken. Duck eggs are excellent for baking, if a little tough fried or boiled.

Show breeds available in the United States include the Call, Black East India, and Crested White. The first-named, in addition to being nice on your pond, attracts wild mallards.

THE YARD

Ducks do not need a pond, despite the fact that they look more at home there. For meat- and egg-producers, a yard equipped with a shallow cement or steel basin full of water suffices. It should be just deep enough for the birds to wade, wallow, and wash their feathers in. Such a man-made pool is best supplied with continuously flowing fresh water.

Sandy soil is best, but not absolutely necessary, for the duck yard. What is necessary is that the soil drain very well. To this end, duck yards are usually located on sloping land. Ducks are quite sloppy and noisy fowl, so don't crowd them. Commercial breeding farms run the birds at five to eight thousand an acre, but this creates immense sanitation problems. Twenty-five to a hundred ducks is plenty for the average farm. That number can be kept comfortable and clean in a 150-by-300-foot yard.

Show ducks on your pond should not be allowed to grow to a flock of more than ten or twenty. If they do, they will trample the banks, destroy your watercress, and generally make a nuisance of themselves. Besides, they'll be competing with the pond fish for nutrients. A few ducks, on the other hand, will give good balance to the pond and increase some of the nutrients.

HOUSING

Housing for ducks is about the same as that for chickens—minus roosts, which they don't use. The two cannot be kept together, however. Mature ducks are cold-hardier than chickens, so their building doesn't have to be as tightly constructed. On the other hand, with warmer quarters than they need, they will develop less protective fat and consume less food, both of which are to your advantage. Provide the house with simple nests set about four to six inches off the floor and comfortably bedded.

The house for young ducklings must have all the litter changed frequently, preferably every two or three days. Because of their drinking habits, ducks throw around a lot of water. Even adding a fresh layer of litter every day, as you must, does not suffice to keep it dry. And wet litter makes for dead or blind ducks: ammonia from the droppings gets in their eyes.

THE FLOCK

Your first time around, buying day-old ducklings is your best bet, particularly if you have already raised a flock of chickens. Brooding and rearing for both are pretty much the same, and you can use the same kind of equipment.

For a duck brood, the room temperature the first week should be 70°F. and the temperature under the hover 90°F. Reduce the latter slowly to 80° to 85°F. the second week, and 75° to 80°F. the third. The fourth through sixth weeks a 70° to 75°F. temperature is fine.

The same encircling guard system as for baby chicks is used to prevent crowding and to keep young ducklings close to the hover. Total brooder area for a hundred ducklings should be around twenty-five square feet for the first two or three days and expanded gradually to one hundred and fifty to two hundred square feet.

If the weather is warm and sunny, the ducklings can have an outside run for a few hours a day as soon as they are three weeks old. But don't let ducks younger than six weeks near the pond, or try to give them water to swim in. They'll get a chill. They are sensitive to not only moist cold, but sun as well. So when they're permitted to go outside, some shade must be provided.

The feed mix for a duck brood is similar to that for chicks. For the first week, feed wet mash from a shallow trough six times a day. Give the ducklings as much as they'll eat; then remove the trough till the next feeding. Grit or sand must always be available. Each feeding should be accompanied by fresh water in a fountain deep enough so the ducks can submerge their bills fully. They need to clean their nostrils of caked mash as well as to drink. Be sure the drinking fountain is not so designed that they can get their whole body wet, however. They'd love to, but even indoors chances are they'd catch a bad chill. The second through eighth weeks, the young ducks should be given a growing ration fed three times daily. Keep the water fountains clean and filled.

When they are six weeks old the brooder can be removed from the house and the ducks allowed to range. At ten weeks they'll weigh in at six pounds or so and are ready for the roaster. This is the prime time for tender ducks, although they are still very tasty if slaughtered a few months later. Ducks are dressed the same way as chickens.

BREEDING

In some ways it's easiest to buy day-old ducklings every year for a new flock. But you may want to keep a breeding flock of your own. If you do, reserve the best of your ducks for breeding stock. Select those with bills, feet, and shanks colored an even, heavy yellow. Look for a solid, compact body, with broad, full breast and short neck. The plumage should be glossy and full, the eyes round, big, and bright. You'll need one drake for every five or six ducks.

Keep a breeding flock separate from the laying flock if you don't gather eggs twice a day. Eggs for hatching should be fresh.

Ducks make poor mothers. So when one of your chicken hens gets broody, let her stay that way instead of trying to break her of it. Confine your ducks one night and the next morning gather as many newly laid duck eggs as you can. If you don't collect enough that day for the size brood you want, you can store the eggs in a cool, damp place—no longer than five days, though. And make sure you turn them twice a day. If not, the old wives' tale goes, the duck will stick to the shell. True or not, fewer eggs will hatch if you don't turn them.

Slip a batch of duck eggs under a setting hen slowly. Remember, a broody hen thinks you're out to swipe her eggs—even if she's sitting on an empty nest. She'll be temperamental and prone to peck. You can allot up to ten duck eggs per broody hen. They should hatch in twenty-eight days, unless you have Muscovy ducks, whose eggs take thirty-four.

A mother hen will leave the nest once a day to eat, for perhaps half an hour. If you're lucky enough to find her away from her post, take the opportunity to sprinkle the duck eggs lightly with warm water, particularly toward the end of the hatching period. Duck eggs, as might be suspected, need more moisture than their land-fowl counterparts. Also be sure to turn the eggs once a day. A chicken will do it herself with her own eggs, but the duck eggs will be a little too big for her to handle.

THE BROOD

Once the eggs hatch, the mother hen should be confined to a floorless cage three by three feet in floor area raised high enough so the young ducklings can crawl under the bottom to roam farther when they want. Hens like to walk, ducks are not as adept at it. By confining the hen, she will not exhaust the ducklings.

Newly hatched ducks do not need to be given food or water the first twenty-four hours. After that, the care and feeding schedule is the same as for your first shipment of day-old ducklings.

When the ducks reach the four-week age, the mother hen may be released from her confinement to guide her ersatz brood wherever she wants. At six or seven weeks the young ducks are ready for a swim. If your ducklings have been raised by a chicken hen, incidentally, be prepared to see her throw a violent fit when the ducks take to the water instead of to the roosts. It may be a few days before she gives up on her rebel swimmers and calms down.

GEESE

Geese are even easier to keep than ducks, for during the green season they can subsist primarily on pasturage. Their range, however, should be separate from that for other livestock, since they are rather sloppy about their hygiene. Also, don't let them pasture in the orchard if you have young trees. They will destroy the bark. On the other hand, they are excellent for weeding strawberry patches; the stupid things much prefer weeds to luscious sun-ripened strawberries. Geese are such weed fiends they've been used for many decades now in "goosing down" cotton, or keeping the fields clean, down Dixie way.

Geese are hardy creatures, and the only shelter they will need is a small house, insulated in the coldest areas, with an open entrance and a floor to keep out the dampness. In the South you really don't need to shelter them at all except to provide shade, either artificial or natural. As an extra plus, the much-neglected goose is very disease- and parasite-resistant.

BREEDS

The geese available in the United States are Toulouse, Embden, African, Chinese, Pilgrim, Sebastopol, Canadian, and Egyptian. The first four are the most common in the United States, and the apprentice farmer should limit his choice to the Toulouse or the Embden. All the others are much smaller, except for the African, and that one is very noisy. The honking might be appealing in the beginning, but you'd soon have more than enough of it. If the Pilgrim is available in your region (which it isn't often), some thought may be given to keeping this breed. Although it matures to little more than half the size of a Toulouse, the male and female have different-colored plumage, which is handy when you're just starting out to breed since it prevents you from accidentally slaughtering the only gander in your flock.

A mature Toulouse weighs between twenty and twenty-five pounds. It is a gray goose with white abdomen sweeping up to the tail. The female will lay an average of twenty-five eggs a year over a period of about a month. However, she can't set on that many. Fifteen will be plenty for the goose. Farm out the rest at the rate of five per broody chicken hen. The number of eggs per hen should not exceed five—you'll see why after checking out the egg size. In fact, a hen is unable to turn the egg, which is necessary, so you'll have to do it for her once a day. Start the chickens setting first, before the goose. In other words, set the first five eggs laid under a hen, the second five under another hen. Then let the goose set on whatever else she lays. Incubation time is about thirty days. When the goose's eggs are hatched, slip the hen-hatched goslings into the nest at night to ensure adoption. Make sure she's taken them in before you leave them on their own. If she hasn't, you'll have to give them back to the hen.

Embden geese are smaller than Toulouse, averaging fifteen to twenty pounds. They are pure white, and thus more popular when feathers are wanted. Although their egg production is lower than that of the Toulouse (about fifteen eggs per goose per year), they will usually do their own setting and they make good mothers.

TENDING THE FLOCK

Geese mate in the fall. One gander will oblige up to four geese, and if you're planning on roasters, bigamy is the best course. Often, however, a gander selects a mate for life. Don't break up the happy couple.

But do keep them away from other ganders during the mating season or some nasty fights will ensue.

As to fights, ganders are naturally pugnacious and, particularly during the rearing season, will often attack anything that approaches a nest. This trait is an excellent reason for selecting Embdens, since with Embdens you won't have to pick up all those extra eggs that may well be guarded by the gander long after the goose has given up on the whole affair. You might think of ganders as only birds, but a couple of good-sized ones can send you to the hospital for a week if they really put their minds to it. Use caution—or stick to ducks.

If you're using a chicken hen to set on Toulouse eggs, she may need watching. The eggs are apt to hatch unevenly, and the hen will stroll off with the first thing that moves, leaving the remaining eggs to rot. So take each gosling from the nest as soon as it's born. If its mother rejects it, or hasn't hatched her batch yet, keep the new gosling in a flannel-lined box located in a warm corner. Keep the box clean and dry.

Once the hen has hatched her five, if the goose rejects the lot, use the same hen-confinement method as with ducks. At three to four weeks, as long as they are fully feathered out, the young goslings may be allowed to swim. Up to this time, however, they must not be allowed to get wet. Don't let them out on the grass until the dew is well burned off for the day. A moist chill is often fatal to bare goslings.

Goslings are raised like ducks, except that they must have fresh greens at all times. Provide plenty of fresh fountain water in constant supply and feed four times a day, as much as they'll take. Chopped, hard-boiled eggs, stale bread soaked in milk just enough to get it moist, and chopped alfalfa, clover, or vegetables with a teaspoon of cod liver oil make a good mix for the first week. The second and third weeks the feed should be a wet mash of cornmeal and chicken growing mash in addition to pasturage. After that the geese can be allowed to fend for themselves on the range.

An acre will carry ten geese. They are destructive to pasture, grazing close to the ground, so rotation is essential. It's recommended that they be given a daily supplemental feeding. A good feed is half cornmeal and half wheat bran or oats, with another 10 percent meat scraps or middlings (wheat germ). Soaking the meal in buttermilk or sour milk to make wet mash is excellent. Water, oyster shells or limestone, and grit should all be available on a demand basis.

In winter increase the grain supplement for geese to 20 percent of their diet and give them legume hay or silage for the rest. They must have roughage. Fresh vegetable tops and parings should be given when available to your geese in preference to your pigs or other farm animals.

Before slaughtering a holiday-dinner goose, put it on a moist-mash fattening ration of yellow cornmeal and oats for a couple of months. Mix equal proportions of each with buttermilk or skim milk to make the mash wet and extra fattening. The wet mash should constitute about 50 to 75 percent of the goose's diet, with the rest pasturage or other green fodder. However, don't switch a goose to a fattening diet overnight or it might develop digestive problems. Switch gradually, over a week, particularly if your geese have been almost exclusively on pasture. In Europe geese are often force-fed to make them extra plump and expand the liver. This practice isn't worth it unless you're a glutton for punishment. Goose eggs, incidentally, make excellent rubber sink stoppers when fried. You're better off hatching them or selling them to someone else who wants to raise geese.

SWANS

If you have a fair-sized pond, a pair of swans, the most graceful of all waterfowl, may well be worth keeping. You'll get neither meat nor eggs from them, but they take next to no care and are esthetically one of the most beautiful of all avians.

Swans mate for life. So buy a pair if you can. Then step back. Male swans are nothing if not ill-tempered. Eventually they will get to know you and come regularly for feeding to supplement their diet of water plants and insects. Even then, it is advised, however, that you make yourself scarce during mating time. The swans will put together a large nest of scraps and twigs. Six to eight greenish-white eggs will be laid and the male will stand guard over them. He is now ready to attack—anything from a cat to an elephant or tractor that approaches the nest. With luck, you'll have some young swans in six weeks. But don't be too disappointed if they die young. Swans often live to sixty years of age—provided they survive their first season.

TURKEYS

There's only one sane word of advice to the beginner on the subject of turkeys. Don't raise them. They are incredibly stupid birds. So much so in fact that if not patiently taught to eat, they won't know how and will starve to death—although once they get the habit, you can't stop them. They won't even learn to drink unless you keep some marbles in the water fountain to give them something interesting to peck at. They are also disease-prone, and have to be brought in out of the rain or they'll catch their death of cold—an exasperating bit of farm routine for the apprentice during the rainy season. A turkey egg omelet *can* be beat—easily. And so can the Thanksgiving turkey dinner. I'm all for tradition. But a modern prepackaged turkey bears no resemblance to the flavorsome fowl of Pilgrim times. Consider the alternative—a plump roast goose.

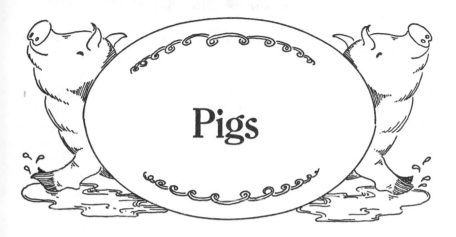

Pigs

Bonn—The German Research Society says the number of pigs dying of heart ailments has doubled in recent years. The society gave no figures, but said apparently the stress of adjusting to modern farming methods is too much for them.

—ASSOCIATED PRESS, JANUARY 29, 1971

THE PIG is the great scavenger of the farm, and no spread is really complete without at least one or two. Besides being fine natural garbage cans, swine are the smartest and, if well cared for, the cleanest of barnyard inhabitants. Believe it or not, pigs do not have to be odoriferously overwhelming. And although you won't want to make your own glue and shoe polish, keeping a couple of pigs on the hoof and one in the smokehouse means you'll be eating well—and so will your land.

At the tail end of the story, pigs will produce sixteen tons of manure a year per thousand pounds of animal weight, as compared with, for instance, eight tons for horses and twelve tons for dairy cattle. And more and more the slogan of the aware farmer is "manure is money." Where once good rich manure was allowed to be eaten by bacteria in settling ponds, it is now almost universally returned to the field. Each ton of hog manure contains five hundred pounds of organic matter, ten pounds of nitrogen, ten pounds of potassium, and five pounds of phosphorus. Think what that can do for your soil. In fact, in China, which has more pigs than any other country in the world, swine have been raised for millenniums primarily for their fertilizer and scavenging properties.

The domestication of the pig began in China as long ago as Neolithic times. Since, however, pigs do not adapt to a nomadic grazing existence the way sheep, cattle, and horses do, they did not accompany the migration of various ethnic groups. Instead we find throughout history different nationalities domesticating their own local wild breeds from scratch. Unlike many other domesticated quadrupeds, moreover, the ancestors of today's American farm swine still roam wild in the forests of Europe—although, of course, in decreased numbers. Their survival may be due partly to the fact that domesticated pigs will revert in only a few generations to the wild state; in fact, they'll mate happily and successfully with their wild cousins if permitted to do so.

There were no pigs in the United States till the arrival of the white man. But because of the hog's transportability—it could survive well enough in cramped ship's quarters and didn't demand fresh forage or exercise—it was among the first food animals to be brought over from Europe. Columbus, who always managed to carry something useful with him either to or from Spain, had hogs on his voyage to the West Indies. Later the mainland of North America was literally invaded by the swine accompanying Hernando De Soto. When he reached Tampa, Florida, he had thirteen head with him. Three years later, at the time of his death on the upper Mississippi, the thirteen had proliferated into a herd of over seven hundred, not counting the strays left behind. These early settlers served as the breeding stock for the North American razorback, in turn a brief source of hunting game for Indian and white man alike.

Transportability and fertility are not the only reasons for the early popularity of pigs. As the farmer soon discovered, in combination with beef or dairy cattle, swine made for maximum economy. If he raised beef cattle, the pigs ate all the food carelessly spilled by the cattle. In dairy production, at least in the past when the emphasis was more on butter and cream, the residual skim milk was profitably used to fatten the swine.

Besides the obvious pork product—meat (easy to cure and delicious smoked)—swine contribute such diverse benefits as insulin (it takes the pancreas glands from 7,500 hogs to produce one ounce), ACTH (over 1,500 pounds of pig pituitary glands to make one pound), glue from the hoofs, high-quality charcoal for specialty steels from the bones, fine brushes, air filters, and padding for mattresses from the hair, shoe polish and buttons from the blood, and on and on.

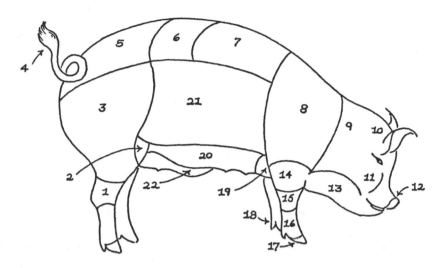

PORK   PRODUCTS

1. HOCK	7. BACK	13. JOWL	19. FORE FLANK
2. HIND FLANK	8. SHOULDER	14. FORELEG	20. BELLY
3. HAM	9. NECK	15. KNEE	21. SIDE
4. TAIL	10. POLL	16. PASTERN	22. SHEATH
5. RUMP	11. CHEEK	17. TOES	
6. LOIN	12. SNOUT	18. DEW CLAW	

If it sounds like by having a couple of pigs around you can eat your bacon and have it too—that's not far from the truth. Pigs have such a position of importance in American agriculture that they were long ago nicknamed "mortgage lifters." Mortgage or no, a few pigs on your farm are almost all gravy.

BREEDS

Swine, because of their short breeding cycle and large litters, have been through more progressive development than any other farm inhabitant. Most swine today originated as crosses between the European wild pig and the Malayan or Philippine wild pig. However, their ancestors would hardly recognize them. The modern hog still has four legs and a snout, but the similarity just about ends there. The wild hog's weight was concentrated up front, with huge head and shoulder development. The farm hog has had its body weight shifted back by selective breeding to produce large, plump hams and meaty bellies for bacon.

The domesticated pig has undergone even more drastic changes in the last century. At one time lard was a primary pork product. For the small farmer, homemade lard is still a great treat, and you get the ingredient with any pig you slaughter, no matter how lean. On a commercial scale, however, massive production facilities for turning out vegetable oils such as soy, safflower, peanut, and corn have sent lard prices skidding to new lows almost yearly. Because of this the traditional lard-type hog has been bred leaner and leaner until it is now considered meat-type. Over 90 percent of the hogs raised in the United States today are of the latter type. The remainder are bacon-type, which, although fattier than meat hogs, are still leaner than the old lard-type.

Since a meat hog will normally produce better hams, you'll probably prefer to start with this type. And even the meatiest of hogs will give you plenty of lard and cracklings, still by-products in meat production, and tasty ones at that.

Duroc (Meat-Type). The Duroc is America's most popular pig, with over four million presently being raised. The breed is named after, of all things, a famous American stallion of the late 1800s, the period when this breed of swine evolved from crossing New Yorks with Jersey Reds.

The animal comes in varying shades of red, from light to dark. Either extreme in coloring is usually shunned in favor of a nice sandstone red. White spots on any part of the body or white feet disqualify the specimen for show (in case you're planning on going hog-wild over this aspect of farm life), as do black spots over two inches in diameter. Mature boars weigh in at an average of 900 pounds, sows at 750 pounds.

Poland China (Meat-Type). This breed comes from neither Poland nor China, but from the Miami Valley area of southeastern Ohio, the biggest corn/hog-producing area in the United States prior to the Civil War. This was the time when Cincinnati acquired the now-forgotten nickname "Porkopolis" because of its extensive packing plants. The basic breeding stock was the Big China breed, introduced into Warren County by the Shakers. A prominent Polish farmer in the valley was supposedly responsible for the first name of the breed that developed when the China was crossed with various local breeds in the valley.

The Poland China is a black hog with six white areas: feet, face, and tail tip. Small white spots are occasionally seen on the body. It has flop ears. Mature boars weigh an average of nine hundred pounds, sows eight hundred pounds.

Hampshire (Meat-Type). The Hampshire is one of the newer and most distinctive in appearance of the swine breeds. It originated in Kentucky and is particularly good for southern farms—in general, dark-colored hogs are preferable in the South, where light ones are apt to be troubled by sun-scald.

The Hampshire is black except for a white belt that sweeps around the shoulders and body down to and including the forelegs. It is a fairly trim hog, with erect ears. Lardiness is low; at the same time, the hams are not as fully developed as in some other breeds. It is a relatively small hog, with mature boars weighing an average of 800 pounds and sows 650 pounds. A good, well-balanced, easy-to-care-for first hog if there are breeders in your vicinity who can supply you with shoats.

American Landrace (Bacon-Type). Throughout history agricultural monopolies have been fiercely protected by nations nature favored with a unique plant or animal—Brazil and its rubber trees, for instance; Holland and its tulips; Denmark and its Landrace swine, producer of the world's finest bacon. In 1950, however, the Danish government permitted the development of purebred Landrace herds in the United States, and the American Landrace was off and running as one of the most popular bacon-type hogs.

For those who always thought pigs were pink and white, here's your breed. Although black skin spots or freckles are relatively common, the hair coat should be pure white. The ears lop, the legs are short, and the body is proportionately longer than that of most breeds.

Yorkshire (Bacon-Type). Another white pig, this time from Yorkshire, England. Descendant of the Old English Hog, this pig by careful inbreeding has been developed into prime stock. The dished face and erect ears give it exactly the look a nonfarmer expects of a pig.

The Yorkshire is excellent for the apprentice farmer who wants to go beyond fattening young pigs bought from others into breeding his own. Yorkshire sows make fine mothers and are well known for the ample supply of milk they produce for their larger-than-average litters.

STARTING YOUR HERD

The best way to learn how to handle a herd of pigs is by acquiring a couple of shoats in the spring. You may find it simplest to just keep buying eight-week-old weaned pigs every spring from a local breeder and fattening them for fall hams. Or you may want to breed your own eventually. It's a matter of money versus labor. If you breed your own, the next year's stock costs you only your labor in growing feed. Under ideal confinement conditions, a professional breeder can get two litters a year from his swine. Your sow shouldn't be made to have more than one a year, but pigs are known for their large litters—five to nine or so. However, it would probably be best to wait till your second year before unleashing a flood of piglets on your farm. Use the first year to get to know pigs. Two will do nicely for a start. The second year, keep four, one of them to breed. The third year, with your sow in the family way, you're set for the pig population explosion on your farm.

KNOW YOUR PIGS

Barrow:	a male pig that was castrated when young
Boar:	the male hog at any age
Farrow:	to give birth to a litter of pigs
Gilt:	the young female pig, less than a year old, that has not farrowed a litter
Hog:	the proper, industry name of the beast
Litter:	all of the pigs born at one farrowing
Pig:	in general usage, the hog, any age, any sex; in technical jargon, a young hog of either sex before it's old enough to go to market
Shoat:	a young, freshly weaned pig
Sow:	the female pig after she has farrowed a litter
Stag:	a male hog castrated after full sexual development
Swine:	hogs collectively

One thing you'll hear frequently discussed among hog farmers is the corn-hog ratio. For the commercial farmer, the corn-hog ratio is an important means of measuring whether at any given time it will pay to raise hogs for marketing. Numerically, it is the number of bushels of corn required to equal in *value* a hundred pounds of live hog. A high corn-hog ratio means cheap corn and costly hogs, a profitable situation for the breeder. Usually a corn-hog ratio of thirteen is considered the break-even point for most producers. For the small farmer raising pigs for his own larder rather than for the market, however, this is not the case. A low corn-hog ratio means corn is dear, but if you're growing your own, that doesn't affect you; it also means young pigs will be cheap, so that's the time to buy, if you can. Perhaps you should even get an extra one or two and plan on eating more pork the coming winter.

There is some lack of uniformity in body conformation within any given breed of hog. Generally, however, for the best shoats check the parent stock for firm, full jowls and a medium neck length. The shoulders should be compact and smooth, leading into a well-arched back whose top is at about the center of the pig. A good, healthy hog rump is divided, with a high-set tail. Look for a firm, straight underline—a middle-aged bulge is no better on a hog than on you and me. Legs should be about a third the total height of the hog. The skin should be smooth, wrinkle-free, and neat.

Now look over the offspring. Pick those of a litter that look obviously healthy and bigger.

Rangy hogs are characterized by excessively long legs. They should be avoided, because the meat is rather thin. You get an oblong pork chop, for instance, instead of a nice plump round one. For just the opposite reason, avoid chuffy, or short-legged, overly plump hogs. Their meat is too lardy. All of this might not seem very important if you're just planning to raise a couple of hogs for home consumption. But the point is a good hog will produce meat more efficiently; you'll end up with more usable pork per amount of feed consumed. And don't be too surprised if after a couple of years of culling and breeding you end up with a hog you just can't resist entering in the county fair.

Weight is an important consideration when choosing either feeder stock for fattening or breeding stock. Shoats should be in the twenty-five-pound range at six weeks of age, the thirty-five-to-forty-pound range when eight weeks old. The parent stock should have reached the two-hundred-pound level at about three and a half months for the sow, three months for the boar—and remember, your tiny, cute piglet is going to do the same.

Since you want to minimize the chance of future problems, check that the herd from which you get your shoats is free from atrophic rhinitis, erysipelas, virus pneumonia, hog cholera, and other diseases. Ask for vaccination and health certificates when buying your pigs.

HOUSING

Commercial hog houses more and more resemble huge mechanical automatons where the pigs are raised in complete confinement. The manure is automatically removed from below the slotted floors of their pens. Food and water are supplied automatically—the constituents of the feed in some cases determined by periodic computer analysis, balancing all the factors of age of the stock, nutriment content and price of available feeds, etc.—and automatic ventilation and complete climate control are becoming bywords in many of the larger "pig parlors," as these confinement houses are called.

Where does this leave the farmer who only wants to raise a few pigs for his own larder? Still in fine shape, actually. The mechanical behemoths of commercial pig farms are geared to the kind of large-scale production where price fluctuations of half a penny per pound in cost come out to thousands of dollars in profit—or loss. All a home-body hog on pasture really needs is an adequate roof over its head with fifteen to twenty square feet of floor space. During cold winters in northern parts of the country, an insulated, heated shed may be necessary. If the pigs are to be kept in their house round the clock, you'll need a minimum of one hundred square feet per pig. This is one reason why for the beginner it's best to buy eight-week-old hogs the first spring to simply fatten up for the fall.

The Portable House. For the small farm a portable hog house is ideal—inexpensive, convenient, and flexible. It is easily built and lends itself well to rotation pasturage, minimizing disease. Or, if an existing small outbuilding is standing empty, that can be used for shelter in the early spring and the pigs later turned out to pasture with nothing more than a shade roof to let them escape direct sun. Even more rudimentary is a large tree supplying the pasture shade; in this case, however, the pigs may well trample down the area directly beneath the tree. The shade roof has the advantage that it can always be moved to avoid stripping the pasture, with the subsequent development of wallows during periods of heavy rain. But do provide one or the other, tree or roof, because shade is essential for hogs. They can't sweat, since they have no glands to speak of for this purpose. Wallowing in mud is a cooling natural compensation for the lack of sweat glands. It's not something pigs do because they are basically dirty animals, as the folklore surrounding them might have us believe.

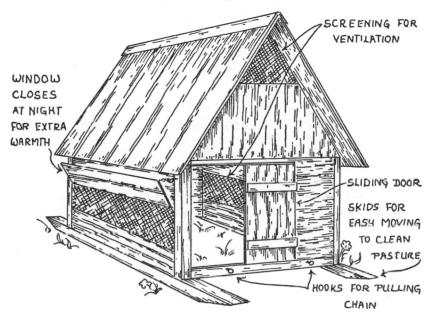

WINDOW CLOSES AT NIGHT FOR EXTRA WARMTH

SCREENING FOR VENTILATION

SLIDING DOOR

SKIDS FOR EASY MOVING TO CLEAN PASTURE

HOOKS FOR PULLING CHAIN

PORTABLE HOG HOUSE

Portable houses and shade roofs can be towed from pasture to pasture as you herd your pigs around. With swine, as with most farm animals, disease and parasites are minimized by rotating on a skipped-year basis the pasture on which the livestock is kept. Better yet, use the pastures on a three-year rotation, keeping the pigs off for two consecutive years.

The Permanent House. If you plan on wintering hogs, in northern climates a permanent house should be provided for at least the coldest months. An unused outbuilding probably can be converted for this purpose, preferably one with a low ceiling to conserve heat. Seven- to eight-foot ceilings are good. Allow about two hundred square feet per sow if you're planning on raising a brood. This permits room for a farrowing pen. The house must be insulated, and the insulation in turn covered by what is known as a vapor barrier. The humidity in a pig house is high—a mature pig gives off almost a gallon of water a day from breath alone—and the warm, moist air will tend to seep out even through the insulation. A vapor barrier impervious to the passage of humid air, simply a layer of aluminum foil (commercial roll-in-place insulation often comes already bonded with it), plastic film, or the like, will keep the steamy air from escaping too quickly. The vapor barrier must be covered in turn with one-inch-thick hardwood lumber so the pigs have no access to it; otherwise they'll add vapor barrier to their menu.

Stale air must not be allowed to remain in the house. There has to be circulation. At the same time, drafts should be avoided. A ready-made wind-powered ventilator can be purchased and installed without much difficulty. Or judicious use of windows will often suffice if it doesn't create drafts. Draft conditions vary from building to building. Walk around in yours and feel for yourself.

Ventilation is necessary, but so is maintaining adequate heat. In a well-insulated house the body heat of the pigs as they huddle together on a comfortably bedded floor will generate enough heat to keep them warm in moderately cold weather. A good winter temperature for the hog house is 50°F. Forty degrees is tolerable, but the pigs will consume more and produce less meat. In a northern region where the outside temperature falls below freezing for long periods of time, you'll probably have to install a heating system.

The sleeping area of the house must be supplied with a good layer of bedding. Materials most commonly used are straw from wheat, oats, or rye, peanut hulls, ground corncobs, or hay. Oat straw is preferred, since it absorbs more ammonia and nitrogen than the others. When using straw or hay, chopping it up will mean better absorption of hog waste, but it can also make it too dusty. The best compromise is to use stalks that have been crushed instead of chopped or left whole. The amount of bedding to use depends on its absorption quality. At a minimum you want to ensure that all urine is absorbed. However, during very cold times you'll have to double or even triple the layer to ensure that the pigs are sufficiently insulated from the cold floor. On an average, you'll need about half a ton of bedding per pig per year.

EQUIPMENT

Assuming you're starting out with just a couple of young pigs in spring, the equipment needed is simple. Besides shelter, you will want a feed trough that's easy to keep clean and designed to prevent the pigs from lying down in it. When you build the feed trough, plan on allowing one and a half feet of length per pig. Use sturdy hardwood, making the flat-bottom trough one and a half feet wide and eight inches deep. Better yet, buy an inexpensive cast-iron one. The trough should sit on a platform or slab of cement to keep the ground from being trampled into a wallow by the pigs. When the trough is light and short, as will be the case with a feeder for two, design the slab or platform so the trough can be bolted down to keep the pigs from pushing it around. The mounting should not be permanent, however. You'll want to remove the trough for an extra-thorough cleaning once a month.

DRUM   WATERING   TROUGH

CUT OLD GASOLINE DRUM IN HALF. SMOOTH OFF ROUGH EDGES WITH FILE. PIVOT-BOLTING TO STAKES AT EACH END KEEPS IT IN PLACE AND MAKES FOR EASY EMPTYING AND CLEANING.

To be healthy, fast weight gainers, your hogs need access to a constant supply of fresh water. Don't let them drink directly from a stream or pond; they'll trample the banks down, destroying areas better reserved for ducks or wildlife. Use another trough, again mounted on a platform or slab like that for the feeder. Count on a half a foot per head with a minimum length of three feet. To supply running water to a pasture trough, lead a flexible plastic pipe, a la garden hose, from the nearest spigot, along the fence line halfway up the posts, to the trough. Better yet, put a Y joint on the hose end and a trough on either side of the fence, and you can supply water to pigs in one pasture and to goats in the adjacent one.

Fencing. "Good fences make good neighbors," and they'll keep the pigs out of your vegetable garden as well. Fences come in all sizes and all materials, from the traditional New England stone "hedge" to the single strand electric. Board or pole fences are fine for enclosing a yard adjacent to a permanent pig house. But for portable pig houses on pasture, your best bet is wire. A combination of woven and barbed is advisable, to prevent your swine from sneaking out from under.

Woven wire is sold on twenty- and forty-rod (a rod is sixteen and a half feet) rolls with a number such as 1047, 939, or 726. The first one or two digits represent the number of horizontal wires, the last two digits the height of the fence in inches. Get a wire with the vertical components six inches apart for your pigs. Most types also come with the spacing at twelve inches; this is for larger animals. Probably not all types will be available in your area, but fencing similar to the 939 or 832 makes a good swine enclosure. One more factor to take into consideration is the gauge, or thickness, of the wire. The heavier the wire, the smaller the gauge number. For instance, a nine-gauge wire is thicker than a fourteen and a half. You can use nine-gauge fencing, but eleven is more economical and still durable enough.

A woven wire fence itself will not keep pigs enclosed. Even if the fencing is right down to the ground, the swine will root their way out somewhere, somehow. To avoid this, keep the bottom of the woven wire fence eight inches from the ground and string barbed wire as the bottom line four inches from the ground. If you think barbed wire is simply barbed wire, you're in for a surprise. There are more kinds of barbed wire around, varying in shape, size, number, and spacing of barbs, than eggs at Easter—so many, in fact, that some western ranchers have regular wall-mounted collections of them, including antiques from as far back as 1874, when barbed wire was first invented. You can use any type for your bottom line. A popular standard size has four-pointed barbs spaced out at five inches. The twelve and a half gauge comes in eighty-rod spools and is about what you need in strength.

Permanent fence posts are being made more and more from concrete and metal. However, costs will probably dictate that you use wooden ones. They're quite serviceable, and durable enough for most purposes. If you cut your own, even if you creosote them, don't be surprised if they sprout. Down the road a couple of years ago there was a lovely fence with willow posts. Last time I looked it was a row of trees. And an avenue of trees for a fence line is not really desirable here, since the growing fence posts will destroy your expensive wire.

The Wallow. If you're farming in the South, with temperatures going up in the 90s for days at a time, you need a wallow or a sprinkling system in addition to your shade roof to help keep the pigs cool. A sprinkler is not economically practical except on a commercial farm of fair size. A wallow can be either a natural mudhole or, for better sanitary conditions, an artificially constructed belly-deep wading pool. One ten by fifteen feet, sloping from a few inches deep at one end to two feet deep at the other, is good. The shade roof should be near the wallow or pool, but not directly over it or your pigs will simply remain relaxing in their tub all day.

FEEDING

The fastest weight gains are made by hogs on a full-feeding schedule. That is to say, the hogs are permitted to make hogs of themselves and eat as much as they want. Simple and efficient as this sounds, it's not always the best method. Boars and breeding sows, for instance, are better kept on different diets. On the assumption that your first pigs at least will be raised for meat purposes rather than for breeding, you can give them all they want any time they want if they are kept in confinement. If they are on pasture, which is not only more economical but easier as well, supplement their diet with a daily grain feeding and a mineral ration, provided in a covered self-feeder so the feed doesn't scatter on windy days or cake in the rain.

Feed mixes vary considerably from area to area and with the time of year. But one of the basics is corn. The Corn Belt is the largest producer of hogs in the United States; corn is close to the ideal feed for pigs, and an excellent synergistic relationship exists in this area between the field crop and the livestock. But in itself corn is not sufficient for a balanced diet.

Protein Supplement. At the turn of the century the awareness developed that pigs, with, like man, only one stomach, were very inefficient compared to their multigastric grazing counterparts, cattle and sheep, in converting low-quality protein to higher-quality protein. Supplementation of protein entered the feed picture and everything changed. Slaughter by-waste that once had been pumped as pollutants into rivers was now saved and processed into a high-protein food supplement given the name tankage. Dried skim milk and buttermilk added impetus to the development of protein-concentrated feeding. These high-quality proteins, all animal-derived, were diversified with the expanded use of blood meal, fish meal, and other fish residue. The amount of the fish residue used, however, must be limited, since overconcentration tends to produce fish-flavored bacon.

Table Scraps. All this, of course, does not preclude feeding your pigs table scraps. Pigs not raised for commercial sale will do fine on garden waste such as cabbage leaves, culled fruit, root-crop tops, and table leavings, particularly if they spend a good deal of their growth period on pasture.

However, a scrap diet should be supplemented with high-protein feed. If you have extra or sour goat milk, questionable eggs, and such, the pigs will make short work of them.

If the scavenger possibilities of your spread don't stretch far enough, local food processors and restaurants can supply additional feed. Garbage was till recently a traditional source of feed for pig farmers. However, the quality—yes, even of garbage—has been declining rapidly, and it now often contains less food and more paper, bits of glass and metal. Also, by law the garbage must be boiled before being given to commercial pigs to eliminate trichina. Since you don't want trichina either, you should follow the same procedure when using restaurant scraps. This often makes garbage feed more trouble than it's worth. Bakeries, on the other hand, are quite often willing to sell you their stale bread at very reasonable prices, and if you're good at bargaining, you'll have a steady source of acceptable pig food.

Feed Mix. You'll probably be feeding your pigs mostly what is excess on your farm. But if you have the time and room, you might want to consider laying down a couple of acres of corn to use as a base for a homemade feed mix. The following is a basic formula for a ton of standard feed.

   1,530 lbs. ground yellow corn
     100 lbs. alfalfa meal
     200 lbs. soybean meal
     120 lbs. tankage
      10 lbs. salt
      10 lbs. limestone
      10 lbs. cod liver oil
      20 lbs. steamed bone meal

Mix well. A balanced formula won't make a balanced diet if the twenty pounds of bone meal are all in the bottom of the barrel. If you pasture your pigs, you can skip the alfalfa meal. In hog-breeding areas of the country basic formulas such as this may be varied almost weekly by computer calculation of feed prices, weight gains, age of stock, etc. If you're just keeping a few head for yourself, such precision is obviously unnecessary.

Ready-Mixed Feed. In some cases it may pay to buy your feed ready-mixed from a dealer. If the idea doesn't offend your sense of self-sufficiency, sit down and calculate seed costs and amount of time and labor needed to produce adequate feed supplies for your herd. You may find it's more economical and efficient in the long run to buy your mix, especially once you start wintering pigs for breeding purposes.

If you're just fattening for fall slaughter, chances are pasturage and scraps, supplemented with protein and minerals, will still be the most economical method. In supplementing, however, avoid too heavy a concentration of soybeans, peanuts, or their by-products. The protein from these is excellent, but they are also high in a type of fat that causes "soft pork." Soft pork is perfectly edible; it just makes for rather mushy slicing of bacon and ham.

Slopping. All right, whether you mix your own, buy commercial feed, or simply use scraps and pasturage, how do you actually go about feeding a pig? (Hog-calling you'll have to learn from a neighbor.) Do you just throw the food in the troughs? Or does it need preparation?

Potatoes and beans always must be cooked to help break them down for easy digestion and full utilization. Don't, however, serve them up steaming hot. Hot food is Western man's particular fixation. It has no place in the hog diet. Certain feed mixes—those containing hard grains, for instance—should be ground before being put out. But not too finely. Dusty food will tend to ball up in the animal's mouth. To avoid this, if you have a good supply of, say, finely ground mill waste available, mix it well with water in an old barrel, the swill barrel. Let the water soak in overnight, stir, and slop it out—hence the familiar phrase "slopping the hogs." All foods can be slopped and, although the trend has been away from this to dry feeding, it now seems to be reversing itself. On large hog spreads a compromise called paste feeding is becoming popular. Paste feed has considerably less water than swill and more than dry. It has several advantages for automated feeding, but if you're going to have fewer than five hundred pigs, it's probably not worth even considering. Stick with slopping. It's the best all-around method for the family pig, whether you grow your own feed or buy it bagged.

Hogging Down. For those who grow their own feed, or have a crop suddenly plastered to the ground, or lodged, by heavy rain or hail, making harvesting impossible, hogging down is useful. If you're raising feed, you save yourself a lot of work; if the crop can't be harvested, you thus avoid wasting it. Hogging down simply means that when the corn, or whatever the crop, is ready to be harvested, you let the pigs into the field. They don't have to be hand-fed; they'll pick their own. Supply them with a mineral-mix side dish as you would pastured stock. Make sure that, also as with pasturage, the mix is kept in a covered self-feeder where it won't be blown away or caked by rain.

Gleaning down is a variant of the same game, except that in this case you let the hogs out in the field after you've harvested the crop. They then polish off the spillage and other unavoidable waste that always accompanies harvesting.

Not only does hogging down save labor, it guarantees that the maximum amount of fertilizer is returned to the field—on the spot. Also, since hogging down is a modification of the pasture system, it tends to minimize disease. Two points are to be taken into consideration, however. (1) You have to either put up additional fencing around the field you're letting the hogs loose in or have someone tend them.

(2) Don't let the hogs on the field when the ground is unusually wet; a large amount of food will be lost due to trampling and the tilth will be lowered, particularly if you have full-grown breeders.

Pasture. For those just keeping a couple of pigs, feeding by the pasture system saves a lot of work. If you're not interested in record-breaking weight gains, your pastured pigs will thrive on just one full feeding of grain daily, a good mineral supplement, and their pasture. Slops and some protein supplement, of course, don't hurt.

A crop of rye in the winter, and crops of alfalfa, soybean, and ladino, red, or alsike clover the rest of the year will give you round-the-year grazing if you're not too far north. A couple of acres of good pasture can usually carry twelve growing pigs or five mature pigs comfortably. But make sure that on the whole the crop is not grazed down to below four inches. If it is, it's an indication your swine haven't enough food around for a comfortable porcine existence. You'll have to expand the pasture and its fencing, decrease your herd, or use more supplemental feed to keep them fat and content. The last alternative, of course, is not economically to your benefit. The best preventative for overgrazing is a well-managed pasture, one that is properly seeded, plowed, limed, and fertilized.

Early spring and late fall pasturing are often on a part-time basis, with the animals alternately grazing a few days and being confined with feed a few days. This practice permits the pasture crop to develop stronger roots in the fall and more vigorous growth in the spring than if the pasture were constantly grazed.

BREEDING

Even on the assumption that your first year or so of pig raising will be devoted only to spring-to-winter fattening projects, either because you have enough to learn already, or because suitable housing for a herd of swine is limited, it still doesn't hurt to know a bit about breeding. Sooner or later you'll want to do your own. And meanwhile, since you're buying the end results of somebody else's efforts, it's helpful to have an idea of what they've been doing.

Hogs may be bred twice a year, in spring and fall. Intensive commercial breeding permits more efficient use of specialized equipment, and in turn lower end-product prices. However, for the apprentice farmer, the older once-a-year spring-breeding pattern is recommended. And it is suggested that you borrow a good boar for breeding purposes rather than keeping your own. It's a lot less trouble and expense.

The Boar. The most important member of a breeding herd is the boar. Since a popular boar may be bred with twenty to forty sows a year, his pedigree, progeny, and meat-producing characteristics influence more pigs than any single sow. Pedigree pays. So does a history of siring large litters.

On farms where hogs are bred on only a part-time basis, however, the boar is often one of the most neglected of animals. The attitude is that he only hangs around for occasional mating sprees and thus doesn't deserve special care. Avoid buying your shoats from a farmer who keeps his boar penned up in cramped quarters with only a make-shift diet, and don't mate your sows with that boar either. A boar should be permitted exercise the year round and have as balanced a diet as his harem. Pastured boars are an indication of good management. One in a filthy small pen is not.

If you see a boar running back and forth along a fence, snapping his jaws viciously and slobbering, don't jump to the conclusion that he's got rabies. His behavior is known as ranting (but not raving). It does not impair his breeding ability, though it is usually subdued whenever possible since it adds nothing to his general well-being.

A ranting boar, incidentally, will give you some idea of why if you have young children on the farm they must be taught that a pig, even a young pink one, is not the lovable little beast they might think it is. Hogs, male or female, young or grown, can be quite friendly. In their own clumsy, exuberant fashion, too much so. For one thing, you never want to make a pet of an animal you're going to eat someday. Sitting down to a Christmas ham can be traumatic if the thought running through your head is "Alas, poor Oscar, I knew him well." More importantly in the case of children, the temptation is always there to hand the family pig something to eat. As far as a pig's intentions go, this is all fine and good, for hogs are far from vicious. But they are voracious eaters and would be hard put to define where the food ends and the hand begins when they are hogging it down. Hand-feeding of a pig by a young child can mean the loss of fingers. Hog jaws are very strong.

Puberty is reached in swine in from four to seven months. A boar should be nine months old and well grown before being put into service. One that is properly cared for may then be mated until he's around seven years old. However, he will not reach his peak capacity until he's over a year old, and will begin to decline in breeding capacity once he's over five.

The Sow.   The sow is not to be underrated, of course. She should be chosen for the same traits as the boar. Add to the list of checkpoints the number of teats she has—don't settle for less than eight, try for twelve, fully functioning—because this is a highly inheritable characteristic, as is her ability to give large quantities of milk.

Gilts often reach maturity before boars of the same breed. Normally a sow is permitted to reach the 225- to 250-pound weight level before she is bred. The body must be well developed and healthy to withstand the nutrient drain of lactation.

Mating.   Sows come into heat on an average cycle of twenty-one days. Older sows usually remain in heat longer than young ones, and the average is two to three days. A sow in heat will mount other sows, urinate more frequently, and often there will be a swelling of and discharge from the vulva. She'll be generally restless and utter occasional loud love grunts. Mate her twice, with a twenty-four-hour break in between, to help ensure fertilization.

Swine usually need—literally—a little outside support when mating. In large-scale production this is provided with the aid of a breeding crate. In more limited operations a couple of bales of hay placed on either side of the sow accomplish the same end. Mature domesticated boars grow to such proportions that it's difficult for the sow to accept service unless she's bolstered to keep her from tipping over.

Pregnancy.   A pregnant sow needs the best of food throughout the average 114 days of the gestation period. This means good pasture whenever possible, and good quantities of ground legumes, particularly alfalfa, during periods when she is confined. The amount of feed and protein and mineral supplements must be well balanced, but is usually limited to keep her from getting too fat. On pasture feeding, supply about two pounds of grain per hundred pounds of pregnant sow, and two-thirds of a pound of protein supplement per sow. Mineral mix should be continuously available. Just before farrowing, the sow's diet is often switched to include as much as 50 percent oats in the grain and a limited amount of linseed oil, both for their laxative properties.

FARROWING

When a sow is ready to drop her litter she'll become extremely restless and begin to make a nest. (You thought maybe only birds built nests?) Milk is often present in the teats before parturition.

In preparation for farrowing, the farrowing house, which should be separate from that for your other pigs, is thoroughly cleaned with a boiling-hot lye solution followed by a thorough boiling rinse. Always handle lye with extreme caution. Fresh, clean bedding must be laid. The temperature of the house should be between 40° and 80°F.; it's recommended that it be kept as close to 65°F. as possible. Infrared heat lamps over the nesting area or farrowing pen will also be needed; newborn pigs are most comfortable when the temperature is 85°F. Make sure the lamps are located so the sow will not accidentally waltz into them. Pig brooders may be used instead, but are not really necessary for the small-scale breeder. Needless to say, the farrowing house should be dry and the ventilation good.

Most modern farrowing pens are equipped with a guard rail that permits the young pigs to step out of range of the mother's hoofs. Almost 50 percent of the infant mortality in pigs can be attributed to the mother rolling over and accidentally crushing the young or stepping on them. Sows are good mothers, but when you're that size and have as many as fifteen little ones trying to tuck in by the udder you're bound to lose sight of one or two when you move around.

As with other firsts in your farm venture, it's a good idea to have a veterinarian or a neighbor with experience standing by to give you a hand with the first delivery. This may mean quite literally lending a hand, incidentally, since if labor goes on for some time without visible progress, it may be necessary to insert the hand into the vulva and gently help things along—a task best performed on the strength of past experience. But usually a sow has no problem giving birth. The piglets come out like little sausages out of a sausage machine.

Contemporary practice dictates the destruction of the afterbirth rather than permitting the sow to eat it. Although nutritionally the modern way might seem inefficient, with today's balanced diet devouring the afterbirth is no longer the important recuperative aid it was for wild pigs. And by removal of the afterbirth, many swine producers are convinced, pig-eating by the sow is discouraged. Dead pigs must, of course, be removed for the same reason.

The umbilical cord of the newborn pigs should be clipped. After clipping, swab down the cord stump with a solution of iodine to prevent infection setting in before it heals.

The four needle, or wolf, teeth—the sharp teeth you find at the front of the jaw in all newborn pigs—should be clipped off at birth, or as soon as you see them. Side cutters make this an easy, painless operation that reduces the chance of the pigs injuring each other or the dam's udder.

During lactation the sow should be fed a diet supplement heavy in protein and minerals. This means more animal protein as well as plant protein than in the normal diet. Pasture will supply the usual good source of vitamins and trace minerals. The sow and her brood can be put to pasture one week after farrowing.

WEANING

Pigs are usually fully weaned when six to eight weeks of age. Two days before weaning, the sow's diet should be decreased 20 percent in volume. Weaning, as with most other animals, consists simply of removing the young from their mother. Make the parting firm and final, keeping the young far enough away from the mother so they cannot see or hear the sow. After the separation the sow should be fed an increased amount of roughage to help dry up the udder.

Starter rations for newly weaned pigs are probably best purchased commercially, since the formula and small quantity often make on-the-farm mixing more trouble and expense than it's worth. When the young pigs reach fifty to seventy-five pounds in weight they can join the regular pasture herd with the same supplemental feeding your other pigs are given.

MAINTENANCE

Branding. In purebred herds the pigs are branded by notching their ears. On a small-scale farm such earmarking isn't necessary unless you're trying to develop improved breeding stock.

Castration. Boars not wanted for breeding should be castrated. Here again, experienced help is advisable. The usual age for castration is three to four weeks, a time when the young pigs are still easily handled and wounds heal readily. If there's no one around to castrate them for you, suckling pig makes a delicious feast.

Tusk Trimming. Although it's not recommended, if of necessity you are forced to keep a breeding boar, you'll want to cut off his tusks each year before he is bred. This would be considered a shocking waste of tusk in the New Hebrides, where boar tusks are carefully trained for years to grow in circles and even double loops; a set of spiral tusks is a valuable medium of exchange. Polishing up a pair for mounting is a fine long winter evening's occupation—it takes forever, as the lack of progress on the pair I brought back from there attest. On the other hand, a boar's tusk that slices your arm down to the bone is not quite so nice. Snub the boar to a post, with a rope around the tusks of his upper jaw, and cut the tusks off with a hack saw. Be careful not to cut the boar— or let him cut you.

PARASITES

Intestinal Parasites. Like all animals, pigs will have parasites, piggyback or otherwise. If they are not kept under control, problems will arise. The most destructive of these are the ascarids, or round-worms. A yellowish-pink parasite dwelling in the small intestines of a hog, the worm reaches a length of half a foot to a foot. Hogs afflicted with roundworms are unable to fully utilize their feed, and may suffer from acute digestive upsets; as a result, they gain weight slowly. Severe cases in young hogs will stunt their growth.

The roundworm lays its eggs in the hog's intestines. The eggs are then passed out with the manure. Incubation occurs while the eggs are on the ground. After being rooted up by the next pig, the eggs hatch, again in the intestines. But they don't stay there. Burrowing through the intestinal walls, they enter the lymphatic system, eventually reaching the liver, heart, and lungs. The worms are then coughed up from the lungs, reswallowed and thus returned to the small intestines, where they mature, lay eggs and start the process all over again.

Adding one ounce of sodium fluoride to six pounds of regular feed makes an effective worming formula. Give the infected hogs as much of the treated feed as they would normally eat in a day. The next day return them to their regular feed.

But the best way to handle roundworms is to avoid them. Rotating pastures goes a long way toward discouraging them. For the most complete worm control, follow the McClean County system of swine sanitation.

McCLEAN COUNTY SWINE SANITATION SYSTEM

1. Sows should be wormed before being bred, but not while pregnant.

2. A well-prepared farrowing pen is one thoroughly scrubbed down with a boiling-hot solution of lye in water before use.

3. A sow should be thoroughly washed with a disinfectant before being placed in the farrowing pen. This eliminates worm eggs as well as germs.

4. The young brood of pigs littered by the clean sow should be allotted sufficient pasture to carry them without crowding.

5. Pasture rotation should be scheduled at least once a year.

6. Rotated pasture should be kept pig-free for two years.

7. Pigs born in hog houses not located on the pasture where they will be raised are best hauled, rather than herded, to the pasture. This lessens the chance of exposure to infection.

External Parasites. The most common hog parasites, lice and mange mites, are nowhere near as destructive as the worms. However, they do make the animal restless; a parasite-pestered pig will jump and roll about in an attempt to eliminate the source of its itching. Overly active hogs burn up more energy, use up more food, and produce less meat than normal. Even if you don't mind the loss to your larder and feed sack, you'll still be happier with clean, content, healthy pigs.

There are a number of chemical delousing agents on the market. However, nondetergent crankcase oil is quite effective; so are safflower, peanut, and other such oils. And you won't have to worry about chemical residues. Lindane, a common chemical delousant, for instance, has been reported to have toxicity problems.

What you want is a greased pig. Get a good stiff brush and a bucket of oil. Confine the pig and give it a good all-over scrub with the oil—a hog usually will love being scrubbed down. Be firm, but not so rough as to cause skin abrasions. Keep the pig confined for two or three hours afterward. Two oil baths ten days apart should eliminate mites and lice for the season.

PIG   OILER

AN OILED BURLAP-COVERED POST HELPS MINIMIZE PARASITES.
DRENCH THE BURLAP WITH NON-DETERGENT MOTOR OIL. SET THE
POST AT AN ANGLE SO THE PIG CAN SCRATCH ITS BACK EASILY.

DISEASES

Atrophic rhinitis. Persistent sneezing in young pigs is often the
first symptom of this disease. By weaning time the pigs' snouts will
begin to wrinkle. This is followed by thickening or bulging of the snout,
twisting of the face, and frequent nosebleeds. Consult your vet. The
usual treatment is isolation and sulfa drugs. Atrophic-rhinitis-free breeding stock is the best preventative. When buying feeder pigs, remember
that those over seventy-five pounds are least likely to contract the
disease.

Brucellosis. This is another disease from which the parent hogs of
your stock should be free. There is no cure for swine brucellosis (re-

lated to Bang's disease in cattle and undulant fever, or brucellosis, in man, and thus contagious to him—although chances are slim you'd ever catch it, and it's more dangerous to the pigs than to you). The symptoms are often hidden. When observable, they include swollen joints, lameness, and swelling of the sexual organ. The end result is usually spontaneous abortion during pregnancy. Testing for brucellosis is by a simple blood agglutination method. Validated brucellosis-free herds usually have been tested twice over a two-month period. Ask for validation if possible.

Hog Cholera. Over the years cholera has been enemy number one of the swine industry. In the old Western movies, when you saw a farmer destroying his herd and burning the carcasses, wiping out years of work and his whole stock to prevent spreading of disease to neighboring farms, anthrax was the killer. Usually the movie producer picked on cattle anthrax because hog movies don't sell—but hog cholera is just as infectious.

Cholera in swine is almost 100 percent fatal, so forget about any cures. Its symptoms are sudden onset of fever, loss of appetite accompanied by weakness, and heavy consumption of water. The animals will walk with a wobbly gait and try to burrow in straw to compensate for chills. Coughing and alternate diarrhea and constipation are also symptomatic. Purplish patches sometimes develop along the belly area. The symptoms are easily confused with those of erysipelas. Call the veterinarian at once.

A strong prevention program against hog cholera has developed in the last decade. Unless there has been a known epidemic in your area the stock should be all right. Some states have eradicated the problem entirely, others are still in the initial stages of the Hog Cholera Eradication Program. If yours is one of these, get a vaccination certificate.

Erysipelas. Another swine disease that can be transmitted to humans, at which point it's called erysipeloid. The symptoms are more or less those of hog cholera, including the purplish bruised-like appearance of the belly. Additionally, edema of the nose, ears, and limbs is often manifest. Again, not a disease to lose sleep over. If cases break out in your area, you'll hear about them soon enough. Meanwhile, it's sufficient to know the symptoms of a sick pig so you can contact the veterinarian if a case ever arises.

Foot-and-Mouth Disease. In case after reading this far you're of the belief that your hogs are bound to come down with one thing or another, here's one you don't have to worry about. It's not around in the United States, and hasn't been since 1929.

Leptospirosis. A newcomer to the catalog of available swine diseases, it did not become a major hazard to hog breeders until the early 1950s. Leptospirosis, characterized by weak, poor-eating hogs, usually manifests itself at farrowing time; the pigs are stillborn or very feeble. Since you'll have someone standing by at the first farrowing, don't worry about it till then.

Clean pasture and good breeding stock will take you a long way from most problems. In general, the thing to look for in the way of trouble is sudden change. If a hog is off its feed, constipated, diarrheic, unusually restless and fretful, or if any lesions or changes in skin texture and quality occur, you may have a problem. Chances are none of these will happen. Still, if you do notice something unusual, call in an expert—that's the only way you'll become one yourself.

READY FOR SLAUGHTER

Slaughtering a chicken poses no insuperable problems for the apprentice farmer. On the other hand, tackling a two-hundred- or three-hundred-pound hog is considerably more cumbersome. Not only that, but—and this is certainly nothing for the city-bred to be ashamed of—if you've never slaughtered an animal before, chances are you'll either be sick or turn the other way when gallons of blood start flowing. A struggling and squealing half-killed hog is dangerous. Besides such torture is inexcusable, since when properly done, slaughtering is almost instantaneous and painless.

For this reason it is suggested that, the first couple of times at least, the pigs be taken to the local abattoir for slaughtering. Pick-up service may be provided in the deal, but in that case prepare yourself for the probability that the pigs will be shot on the spot and loaded up in carcass form. If you're doing the hauling yourself, you're going to have to get a two-hundred-pound pig up onto the bed of your pickup. Lifting is obviously not the solution. A pig can be made to walk up a plain plank for what he considers no reason at all, but plan on spending the better part of a day convincing him to humor you. The secret of enticing him aboard is food. Throw a few ears of corn into the truck, let the tailgate down and he'll scramble for it. The pig may realize he's going up, but so intent is he on dinner that he hardly sees the dizzy heights from which he might fall. A neighbor may have a stock ramp he'll lend you; if so, by all means obtain it, as it simplifies the entire procedure considerably.

Arrangements with someone else to do the slaughtering and butchering can be either for cash or for barter—part of your pig for his labor. The first time, watch and learn. The second time, do your own butchering. The third time, you may want to do your own slaughtering too. But be sure you're capable of handling it. It's nothing for a half-hearted try.

Besides, you'll have plenty else to keep you busy—getting ready to make that fresh hickory-smoked ham, those spicy sausages, crisp cracklings, and tasty lard that bears no resemblance whatsoever to those square chunks of library paste in the supermarket cooler.

Honeybees

*Who deals with honey
will sometimes be licking his fingers.*
—OLD FRENCH PROVERB

A NATURAL SOURCE OF SWEETS is something you'll probably want on your self-sufficient farm. For those far enough South, there's sugarcane. For those in the North, there's sugar beets. But these are a lot of work to harvest, extract, and prepare. Why not just let nature do the job?

Bees will gather, purify, and store your sweets for you. They'll do everything but put it on your table in a jar, all with little care on your part. And on the side they'll provide you with beeswax for candles and pollinate your crops.

Beekeeping and chicken-raising go well together on the farm. Bees require the least care in winter and early spring, when the chickens most need looking after. In summer and fall, when you're busy tending the swarms and gathering honey for your larder, the chickens are practically taking care of themselves. Chickens can peck away among beehives without being bothered by the bees. They'll eat grubs and other insects that might harm the hived denizens, while leaving the bees themselves alone.

Then too, a natural synergistic relationship exists between bees and an orchard. The bees glean nectar and pollen from the blossoms, at the same time pollinating from stamen to pistil and flower to flower as they make their busy rounds. This increases the setting of fruit considerably. The same holds true for clover and other field crops. Fruit growers who don't themselves keep bees often rent hives from apiaries for just this reason when their trees are in bloom. The beekeeper is paid for the use of his bees—and gets the honey to boot.

If you keep, say, five to ten hives, you're certainly not going to live off the proceeds. But you will have a steady supply of sweets, more bountiful crops, and maybe a bit of extra money to put toward farm equipment you can't make yourself. Even keeping just one hive will supply more than enough honey for the average family.

BREEDS

Wild Bees. What kind of bees to have around is a question worth a little thought. You can go for a walk in the woods, keeping a sharp lookout for a bee, of course. When you spot one, follow it around. Once you've both gamboled among the clover awhile, and the bee has finally led you to the tree wherein his colony resides, you can smoke them out, capture the queen, and carry the swarm home to your hive. It's certainly the romantic way. But not the practical one. Not the least of your problems is the viciousness quite often characteristic of wild bees. Besides, they're usually inefficient about making honey, for all the talk about organized bee society, and disease is often rampant in a wild hive. They're bound to give you no end of trouble, swarming and generally making nuisances of themselves. You're better off with domesticated bees.

Italian. Italian bees are the most widespread commercial honey producers in the United States, and the most readily available for starting new colonies. Partly this is due to their early introduction on the American scene. But of equal importance is the fact that they're gentle, hardy, good workers, and not inclined to swarm irregularly. They are considered esthetically the most beautiful of bees. Two or more yellow bands circle their bodies. Those with two are newly imported immigrants; those with three or sometimes more (up to five have been seen) are domestically bred. The domestic Italian bee is an excellent choice for the beginner.

Caucasian. These bees come in two separate groups: those bred from parent stock originating in the Caucasus Mountains and those descended from valley-bred bees of the same region. The mountain-bred are among the gentlest of all bees and thus very suitable for the beginner. Their quiet habits are such that they are the bees most often kept in small villages or towns, where neighbors tend to look a bit askance at a beehive. However, if you buy Caucasians, make certain the bees you get are the gentle mountain version; their valley-bred counterparts have vile tempers. Caucasians are the best bees for beginners as long as, and this is essential, you get your colony a new queen every year. If you don't requeen yearly, and the old queen dies, the colony is apt to breed itself into a crossbreed with the vicious characteristics of valley Caucasians.

Other. There are bees from just about everywhere; for instance, the Carniolans, native to the Austrian Alps, Cyprians, Syrians, and Egyptians, as well as hybrids such as the Starline and the Midnite. Until you become a dedicated apiculturist, however, it would be wisest to stick with either the Italian or Caucasian variety. Not only because of their gentleness, but also because of their disease resistance. Personally, I'd recommend the Caucasians, particularly on southern farms where Italians tend to crossbreed with the local black bees, producing a nervous, stinging variety.

MEMBERS OF THE BEE FAMILY

With most "livestock" you've got only males and females. When it comes to bees, you have three different groups to keep straight: the queen, the drones, and the workers.

Queen. The queen bee is an egg factory. Her only job is to lay the eggs that produce the rest of the colony. As such, she lays two kinds of eggs, one that produces drones and another that evolves into workers. And she does it with a fury. A queen on an average day will lay fifteen hundred eggs. Not all will develop into brood, however.

In consideration of her function, the queen bee's body is considerably longer than that of the drone or the worker. Her wings, on the other hand, are smaller in proportion to her body. She possesses a sting that she never uses (you can even squeeze her lightly between your fingers and she won't sting) except on other queens, which she strikes at once. Hence there is only one queen per colony. And as the queen goes, so goes the colony. This is why it is important to check on the condition of your queen occasionally, and why you must "requeen."

Requeening is a yearly affair. Bring a fresh queen to the hive in the fall and she will produce a large brood the following spring, ensuring you plenty of bees in time to harvest the nectar.

Drone. Like the queen, the drone has but one function. His is to fertilize the queen. But he doesn't do this continuously. One nuptial flight—mating always takes place outside the hive—and it's all over. He, perhaps a hopeless romantic, dies immediately after fulfilling this sole function.

The drone develops from an infertile egg, as opposed to a fertile one. The cell from which he hatches is slightly larger than that of the worker, although smaller than that of a queen. The drone is absolutely useless to the hive except in his reproductive capacity. He can't gather nectar or pollen or produce wax. He doesn't even have a sting. He lives off the hive till his day of sacrifice.

Because of their relative uselessness, drones should be limited in any well-managed hive. The bees themselves know this. Although hundreds of drones may be permitted to hang around the hive till mating season, once it is over they're shoved out the front door, and starve to death the same day.

Worker. The worker bee is a female that never reaches queenly status. Workers are by far the majority members of the hive. As such they rule the colony. The labor class here dictates the queen's life and decrees the banishment of the drones when the time comes.

The mature worker is a field bee, gathering nectar and changing it into invert sugar. (This is why honey is better for you than sugar. With sugar, your body has to do the work. With honey, the bees have already predigested it for you.) The worker also has a small cavity on each hind leg, used for gathering and storing pollen till it can be brought back to the hive and converted to pap, the formula fed the larvae, and a variation on the menu for the mature bees. Propolis, the glue used by bees for holding things together around the hive, is also procured by the workers from natural plant gums and resins.

Workers do not live long. In the spring and summer main pollen flow their life span is only four or five weeks, at the end of which they die from overwork—some actually wear out their wings. During the winter season, when things are cozy and there's little to do, they may live as long as six months. But come gathering time again, those that have wintered will work themselves to death as well. This is why it is

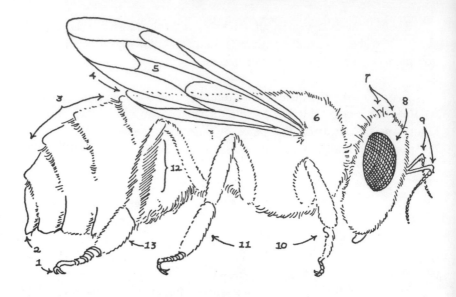

WORKER HONEY BEE

1. CLAWS	6. THORAX	11. MIDDLE LEG
2. STINGING APPARATUS	7. SIMPLE EYES	12. POLLEN BASKET
3. ABDOMEN	8. COMPOUND EYE	13. HIND LEG
4. HIND WING	9. ANTENNAE	
5. FRONT WING	10. FRONT LEG	

essential to have a good queen. The slaves must be continually replaced or honey production ceases.

STARTING A BEE COLONY

If you're not going wild-bee gathering in the clover, impractical at best, the two ways of getting a colony started are to buy an existing one from a local beekeeper or to order a couple of pounds of bees by mail. (There are three thousand to thirty-five hundred bees per pound, in case you don't care to count them.)

Purchasing a colony from a neighboring apiary has much to be said for it. Just make sure the bees have been recently inspected and come with a certificate of health. Once they're yours, they will also have to be checked out regularly by the county bee inspector, so notify him when the colony moves in. Bee inspection might seem an unwarranted intrusion and a general nuisance. It is not. Inspection is of the essence in maintaining a healthy bee population throughout an area. Without it, both you and your neighbors would have nothing but trouble instead of honey.

Assuming you get your colony from a nearby apiarist, your choice of breed will depend on what he is raising. Chances are his bees will be Caucasians or Italians. Either is fine, as long as you remember that Caucasians should be requeened yearly.

A real advantage of buying locally is that the colony will come complete with lots of chatty information. Beekeepers love to discuss their various methods, and you'll no doubt be getting plenty of tutorial apiarian lessons. The neighbor supplier will also probably help you bring the colony home—in the evening when it's dark, assuring that the workers have all returned to the beehive—and set up your hive. This is a gentle way of being introduced to the company of ten to fifty thousand bees. A somewhat nerve-racking experience if the only other bee you'd ever seen was a plastic one in some lady's bonnet.

If there is no apiary nearby that can supply you with a starter colony, you still don't have to head for the woods in search of a stray bee. Shipments of packaged bees crisscross the country every spring. There are numerous suppliers able to send you a few pounds. Most of them are located in the South, where climatic conditions permit year-round development of bees.

Bees commonly come in packages of from two to five pounds. The best starter package is three pounds. This gives you a colony large enough to gather nectar in time if the flow in your area should have a short peak, as well as time to build itself up to a healthy fifty to sixty thousand inhabitants without outside help. Packaged bees are shipped with a queen bee enclosed in a small, suspended, private cage, as befits royalty. There is also a punctured container of sugar syrup from which the bees may dine en route.

When to have packaged bees delivered depends on your location. As soon as the queen is made comfortable in her new hive, she'll start laying. Young bees hatch three weeks after the eggs are laid. Two weeks later they are mature enough to gather nectar. All in all, it takes about eight weeks from arrival for a colony to build itself up to a good harvesting capacity. With this in mind, you should get the bees in enough time so the colony will have itself a sizable labor force by the time the main nectar flow starts. But you don't want them arriving too early. At a latitude of about Virginia, the middle of April is prime colony-starting time. The end of April or the first week in May is usually more suitable in northern climes.

Since you can specify desired time of arrival, order your packet of bees early. You want to time things so the bees arrive when the fruit trees are just beginning to blossom because they will need the pollen to raise the brood. The nectar for honey they can still get later from basswood, wild flowers, and legumes such as clover and trefoil, if the orchard has run out of blossoms by the time the brood is raised to working age.

APIARY LOCATION

There are four primary objectives in planning where to put your hives. First, you need shade. Direct sun beating down on a hive all day is bad for it, particularly in the South. Two or three hours of shade during the hottest part of the day are ideal.

The second consideration is accessibility of nectar-producing plants. Your orchard, no matter how small, is an excellent setting for an apiary. Not only do you have blossoms for nectar, blossoms which will in turn develop into more fruit because of the bees' volunteer pollination efforts, but the trees solve the shade problem as well. Don't locate a hive directly under a tree, however, since some sunlight is desirable. If your orchard is small, it's a good idea to put the hives on the edge nearest to a meadow or a field of clover (except for red clover, whose nectar bees can't reach), raspberries, buckwheat, goldenrod, or other nectar-producing plants.

In northern sections of the country, particularly, your third concern is wind. The hive must be protected from strong, cold winds. A windbreak hedge is fine; a slotted board fence is good if there is no natural shelter. Should your orchard consist of a dozen trees or so, placing the hive in the center will provide sufficient protection in most areas.

The last consideration is simple enough. Point the entrance of the beehive away from any frequently used paths. A heavily laden bee comes in like the Red Baron with his engine on fire. Sometimes it makes a crash landing on the cleat by the entranceway. Sometimes it doesn't even quite make it that far. Whatever the case, bees coming home on a supply run are in a rush getting there, and they don't like bumping into people on the way. They might even sting. It's worthwhile to locate the entrance so you don't pass between it and the nectar-gathering grounds too often while doing the daily chores.

THE BEEHIVE

Home is the bee, home from the glade. And home is a hive. The modern hive, a model of efficiency that has sent the old straw skep and hollow-log gums to the museum, is humane, clean, and safe. The combs are easily accessible. They can be removed for emptying without killing the bees—a feat impossible with the old-fashioned hives, for all their esthetic charm—and are just as easily replaced. Gone are the days when a whole colony had to be sacrificed for a batch of honey.

Essentially, the modern hive is a series of boxes divided to hold frames. On these frames the bees lay down combs to fill with honey. When they've filled the frames, the bees are "excluded" and the frames removed. More boxes can be added to the series; supers are extra frame-holders used when the honey flow is at its peak. Expanding the hive cuts down on swarming. If a hive gets too crowded, the workers are apt to produce another queen, and half the colony takes off with the new royalty, a rather hectic departure known as swarming. When that happens, if you don't want to lose the bees, you've got to track them down and coax the colony back. By adding new supers onto the hive, you get them to stay home, and have more honey as well.

Although it's often possible to buy a hive from a local beekeeper, it's best for the beginner to start with a brand new one. It will cost a little more, but it also assures you of starting with disease-free quarters for your colony, and assembling it will give you an understanding of the way it works that you'll never get from a preassembled unit.

The Beginner's Kit. A basic beehive can be bought as a beginner's kit from bee equipment suppliers. The kit usually includes a standard hive, comb foundation, and frames, complete with assembly instructions. You can put a starter kit together in an afternoon. Basic equipment—a bee veil, gloves, boardman feeder, smoker, and hive tool—also comes with the kit.

Honeycombing. Once you've put the hive together, your next step in preparation for the bees' arrival is filling the frames with comb foundation. Bees are instinctively some of the best architects around. For centuries they have used the hexagon as the basic shape, and this shape can't be improved upon when it comes to honey storage. Unfortunately, bees left to their own devices will not honeycomb the frames properly end to end, but rather will curve the layers of comb, for additional strength. Also they'll build in a fair number of drone cells.

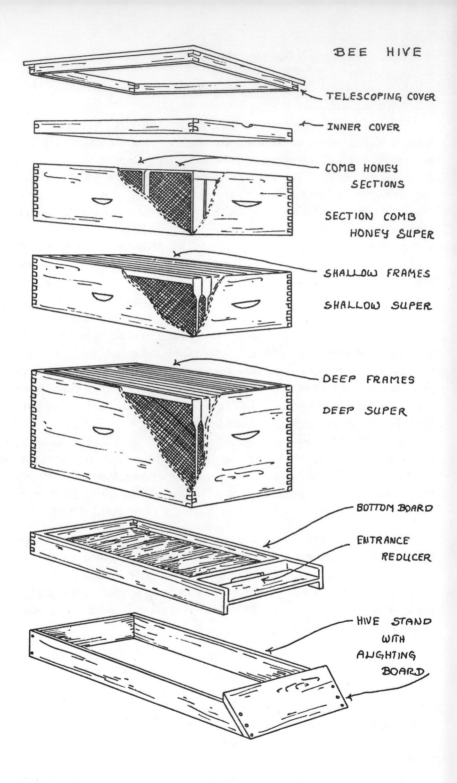

BEE HIVE

TELESCOPING COVER

INNER COVER

COMB HONEY SECTIONS

SECTION COMB HONEY SUPER

SHALLOW FRAMES

SHALLOW SUPER

DEEP FRAMES

DEEP SUPER

BOTTOM BOARD

ENTRANCE REDUCER

HIVE STAND WITH ALIGHTING BOARD

In nature both these approaches are necessary for perpetuation of the colony. On a farm they only reduce honey production. To overcome these two problems, sheet-comb foundation is installed in each individual frame before its insertion into the hive.

As its name implies, comb foundation is a sheet of partially finished beeswax combs which the bees will figure someone else in the hive has already started, and so decide to complete the job. Besides assuring readily accessible combs and leaving room for only a minimum number of drone cells, the foundation sheet permits the bees to devote more time to making honey, since part of the comb is already built for them. When the honey flow is heavy, particularly, the bees can just keep on filling comb instead of having to take time out to build complete new cells. In nature bees do not plan ahead and set up large spaces of comb reserve; rather they build cells as they need them.

The reason bees don't naturally work on the "build first, fill later" principle is that comb left empty would tend to sag, especially in the warm season of prime honey flow. To compensate for this tendency, comb foundation is made extra strong. Support is added in the form of crinkly vertical wires embedded in the wax sheets you buy. This is known as prewired foundation. A less effective sag preventative is three-ply foundation, a pure beeswax comb pressed from three separate sheets rather than one, giving it the same strength advantage as plywood.

To hold comb foundation in place within a frame, thin wire, usually No. 28 tinned, is stretched in rows horizontally across the frame. The wires are then embedded in the foundation. All the intricacies of wire spacing and so forth will already be taken care of in the frame you buy, since it will have holes located in the appropriate places. However, you will want to make an embedding board. This piece of equipment is simply a piece of board that can be fitted snugly within the frame to support the wax and wires during the embedding process. You will also need a wire embedder, a small, handled spur wheel that when heated presses the wire into the foundation gently but firmly.

Traffic Control. Part of efficient hive management is traffic control. If you want comb honey, which is harvested and served up still in the comb, as opposed to extracted honey, you will need to keep the queen bee isolated in the bottom part of the hive. Otherwise you will have brood developing in the honeycomb you wish to use. To keep a queen bee in her place, a queen excluder may have to be used. This is a frame separating the hive into queen and no-queen areas. The frame has parallel wires strung so that the space between them is large enough to let the workers through while blocking the bigger queen's movements. However, if you're not making comb honey—and it's advised that you don't the first year—forget about the excluder, since it will only cause unnecessary traffic jams. Use brood chamber reversal, and you can keep the queen out of your honey supers enough to avoid any problems with extracted honey.

For brood chamber reversal, two chambers are used. When the bottom one is full of hatching eggs, it is placed on top, with the empty one formerly on top put on the bottom. Now remember, these are brood chambers, not the supers from which you get your honey. The honey supers will be even higher up on your incipient skyscraper. With brood chamber reversal, the queen finds an empty chamber right above her, and she keeps on laying eggs there, while the eggs she laid before are hatching—unbeknownst to her, and through one quicker-than-the-bee's-eye movement of your hand above her. Since she has the empty brood chamber handy, she won't bother to climb up into your honey supers.

Another ready means of traffic control is a one-way device that has been developed. When you remove supers full of honey, you don't want them full of bees too. Yet normally they always are. To alter this, an inner cover, or ceiling, is placed between the desired full supers on top and the rest of them still being filled below. The cover is solid except for a bee escape in the center, which has in it a pair of flexible brass strips coming together in a V. The bee pushes its way into the mouth of the V, at which point the sensitive brass strips open up enough at the end of the V to let the bee through. A bee trying to get back in, however, would have to force the V apart at its point, which it can't do. Once the inner cover with its bee escape is placed in a hive, all the bees in the supers will make their way down into the hive proper, where the hive entrance is located, within forty-eight hours. You can then remove the evacuated supers to get at the honey.

EQUIPMENT

There are several pieces of apiarian equipment that may or may not be used by the professional beekeeper. For the beginner, however, they are indispensable, if only because they give added confidence when dealing with bees the first time. Having a few thousand bees buzzing around you can be an unnerving experience. And the strange thing is that a perfectly well-behaved swarm that's never bothered anyone may suddenly decide to go on a stinging spree when there's someone frightened around. As I was told when I was a trembling kid approaching my first beehive, "Don't be scared; if you are they'll smell the sweat and sting."

The bee veil is the first thing you'll want. It is made of either metal screening or the more traditional fine mesh mosquito netting or grenadine. Originally the nets were of tulle; today nylon fiber is often used. The mesh net is draped over a very wide-brimmed hat to keep it well away from the face. It hangs as far as the chest or waist, where it is snugly tied to assure nonentry by bees. The metal screens are usually self-supporting, tying below the neck by means of an attached skirt. For total coverage the bee veil may top off a cool white bee suit, enclosing you in a total bee-proof mini-environment.

Whether you use just the veil or the whole outfit, gloves are recommended. Usually these come in white leather or canvas. They are elbow-length, with drawstrings.

Primitive man used to smoke bees out of their trees with torches. Today's method is the same, if more refined. Smoke subdues bees when properly applied and not in overwhelming quantities. For this reason a bee smoker is one of the apiarist's most useful tools. Basically it is a small, cylindrical firebox in which greasy waste material, rotten wood, or anything else that will make smoke, is allowed to smolder. A nozzle points off the front, directing the smoke where you aim it. Attached bellows are operated with the holding hand, leaving the other hand free for dealing with the bees. The bellows force air into the bottom of the firebox and up through the smoldering fuel, keeping it alight.

You'll need one more piece of equipment to properly get at the honey: a hive tool. A plain steel strip about ten inches long, curved over at one end, the hive tool is indispensable for prying off covers, removing frames, and scraping off propolis gum deposited by the bees in unwanted places. Propolis once served an important function in sealing cracks in tree-trunk hives, protecting the inhabitants from the elements and enemies such as ants. In the modern beehive it serves only to glue otherwise movable parts, such as the frames, together. To separate them you need the hive tool. A screwdriver will work, but not nearly as well, and a hive tool costs less than a couple of pounds of honey.

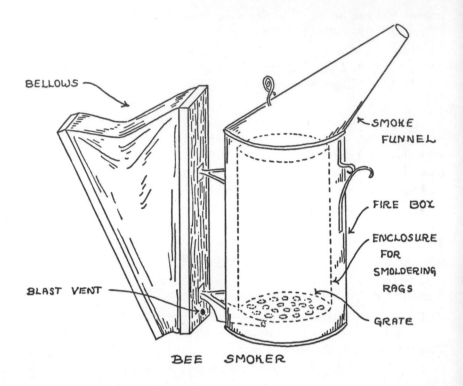

BELLOWS

SMOKE
FUNNEL

FIRE BOX

ENCLOSURE
FOR
SMOLDERING
RAGS

BLAST VENT

GRATE

BEE SMOKER

ARRIVAL OF THE BEES

Contented bees are easier to handle than unhappy ones. In welcoming a colony of them to your farm, your first job is to put them at their ease. If they were shipped, place the bee package in a cool, dark spot where the temperature is below 70°F. Sixty to 65°F. is excellent. Leave them alone for an hour. Shipping is not always as gentle as it should be, and a little rest will help them forget about a rough journey. While you're waiting, mix up a solution of two parts sugar to one part hot water. The sugar will take some stirring before it all dissolves. Smear the syrup liberally onto the wire cage so the bees can stuff themselves on it. Let them lead the life of plenty all day. The best time to install bees is at dusk anyway. When it's half-dark the bees are more likely to settle into a hive at once.

Meanwhile, prepare your hive, already at its permanent location, for the influx. Take off the cover and put an empty super, without frames, on top so that the shipping crate can be set inside it and covered over. A piece of stiff canvas placed over the shipping cage will ensure that the bees leaving it descend, rather than staying in the empty super and starting to make combs there—or abandoning the hive altogether. After all, an empty super is pretty much like a hollow log. You have to force them down so they'll discover the beautiful comb foundations you have ready for them.

The hive entrance should be stuffed loosely with a little fresh grass. It will wilt, and the bees will remove it easily enough by the following day. You'll want to check on it then, to be sure the grass tuft has been pushed aside. Now you're ready.

The next step will probably be a first-class frightening experience, but if the bees are properly gorged there's little actually to be afraid of. And, of course, you're wearing your gloves and veil for the sake of confidence, maybe even bee overalls. Experienced beekeepers do it in shirt-sleeves. But save the bravado for the next time around.

No smoke should be used when installing packaged bees—you want them as undistracted as possible. Stick a long, thin nail in your belt where you can get at it easily, pick up your bundle of bees, and march them to the hive. Tip the bee cage slowly to one side until the top with the feeder can is facing down at an angle 45 degrees to the ground. Lift the cage till you have it right over the hive. Slide the can out slowly while lowering the cage onto the top edge of the hive. After removing the feeder can, take the queen cage out of the bigger cage. Make sure she's alive—if she isn't, reinsert the feeder can and notify the shipper to rush you a new queen. At one end of the queen cage there will be a perforated lid. Remove this, exposing the "candy." The candy, a mixture of honey and confectioner's sugar, is not unlike the Turkish taffy you used to munch on during grade-school math class. Here it serves as a door. The queen is liberated when the workers have nibbled away the door. Take the nail from your belt and puncture the candy. This will help the workers liberate the queen. Hang the queen cage, candy end up, between the middle frames of the super just below the one holding the shipping cage. Invert the shipping cage over the spot where the queen is suspended. The bees will flow out of their cage and toward the queen.

Since, if you're starting from scratch, there will be no honey available for the bees, you'll have to supply the initial feed. Refill the feeder can with another warm sugar solution. Wrap it in burlap to help keep it warm. But be sure to leave the holes free so the bees can get at the "honey." Invert the feeder next to the bee cage. Then cover the bee cage and feeder can with the piece of stiff canvas, draping it down so the bees cannot remain in the empty super. This will also prevent the raiding of your hive by bees from strange colonies, since the canvas should cover all of the exposed frames.

On the second day after your bees' arrival take out the shipping cage, but leave the canvas cover. Save the shipping cage, by the way; it will come in handy at requeening time. Check to see that the grass tuft at the hive entrance is gone. A boardman feeder, on top of which you invert a mason jar, should now be placed in the bee entrance and used to feed syrup. Install it the second or third day, but keep the feeder can full too. Do another syrup check on the sixth day.

On the sixth or seventh day, sometime between ten and two o'clock when it's warm, check to see that the queen is laying. Pick a clear, pleasant day. If you have to wait a day or even two to avoid cloudy skies or rain, it's still well worth it. Bad weather makes for bad bee tempers. The queen will be quite easy to spot, since during egg-laying her abdomen is swollen with eggs. Usually your first indication of where the queen is located is a ring of workers paying court to her, feeding her, washing her, and attending to all the other courtesies to which royalty is entitled. In truth, of course, she's just an egg machine. The cells in her vicinity will be both capped and uncapped; that is to say, covered with a layer of wax or not. Check the uncapped ones for little white eggs. Their presence indicates that everything is all right with the colony and the queen is indeed laying eggs. Leave her to her business. Laying fifteen hundred to two thousand eggs a day is a full-time job.

Remove the syrup can, canvas, and empty super at the end of the first week. Retop the hive with its inner and outer cover. Now your colony is officially in residence—hopefully.

Introducing bees into a new hive is sometimes followed by swarming. The bees, led by the queen, simply decide to find different quarters. The reason for their behavior is uncertain, since the modern hive simulates almost ideal conditions for the colony. Maybe, anthropomorphically speaking, they just want to get out of sterile suburban prefabs and into something more natural.

BEE HANDLING

Bees can get along with as much or as little care as you want to give them. Some keepers are always puttering around the hives. Others only clean them up in spring, empty the honeycombs in fall, and winterize the hives when necessary. Whether you fall into one of these extremes or not, you're going to have to handle the bees occasionally.

The first thing to learn about handling bees is, of course, how not to get stung. In this line the following points are important.

1. Always choose a warm day to work on the hives. Cold bees are nasty bees.

2. The best time is between ten o'clock in the morning and two o'clock in the afternoon.

3. Rainstorms affect the bees' temper. Always let a day intervene after heavy rain before working with your bees.

When preparing to invade a hive, stoke up your bee smoker. Get some oily rags, dry rotten wood, or other material that will burn while producing good smoke. Make sure whatever you use is going to stay lit; you don't want it to go out in the midst of introducing yourself to some fifty thousand possibly stinging bees. The material should smolder slowly, without, of course, starting a major blaze. Smoke is of the essence. Give an occasional puff with the bellows to keep things glowing, but don't overdo.

Wear loose, light-colored clothes. An old college sweat shirt is good, but make sure you've got it on with the fuzzy side in. Tuck your pants into your socks. Sweat suits often have drawstrings at the ankles; they're ideal. Put on your veil and long-sleeved gloves. Taking the smoker and hive tool, approach the hive from the rear. Standing clear of the flight pattern for the entrance, puff a few smoke clouds into the doorway to subdue the guards. It's not necessary to say "This is a hold-up." Be subtle, calm, and don't overdo the smoke or you'll start a stampede. You want to subdue the bees, not set off a fire alarm.

Lift off the top cover. Do it smoothly so as not to jar the colony. Next, with the hive tool, lift the inner cover a fraction of an inch. The crack should be narrow enough so bees can't squeeze through it. All you want to do is let them have it with two or three puffs of smoke. Slowly. Gently. Wait a minute. Puff in two or three more times. This should drive the bees down between the frames. Now lift the cover off entirely.

While you're inspecting the hive, use the smoker whenever a lot of heads begin to show at the top. Smoke them down. Early in your colony's development the frames probably won't be stuck together with propolis. If they are, however, you'll have to pry them loose before you can lift them out for a close look at how the honey factory is doing.

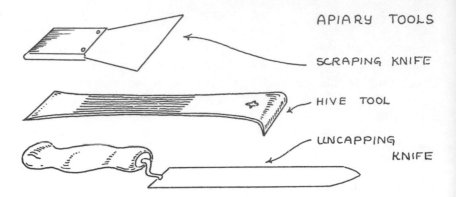

APIARY TOOLS

SCRAPING KNIFE

HIVE TOOL

UNCAPPING KNIFE

Avoid jarring movements and keep puffing in smoke whenever necessary. Use the hive tool to push the frames gently away from both ends of the super, crowding them together. Then push a single frame back toward the end again and you'll have enough space to get your fingers around it. Remember, wearing gloves is going to make you a little clumsier than normal. Go slowly. Lift the frame by both ends. It will be covered with bees, so be sure not to squash or pinch any with your fingers or against the adjoining frames. If there are too many bees just where you want to handle the frame, you can push them aside gently with a gloved hand without upsetting them.

Keep your cool. Bees do an awful lot of crawling around. This is natural. However, if you notice their bodies twitching nervously, or if they line up in long rows head-to-head between the tops of the frames, chances are their mood is foul and they're sizing you up for an attack. In that case, it would be best for the apprentice to close up the hive and wait till the next day. A "sting-a-second hive" one day will be so gentle another day that you'd think it was two different colonies. Closing the hive is just the reverse of opening it, with less smoking.

THE STING

If you do get stung, brush or scrape off the stinger at once with the edge of the hive tool or a knife blade. A stinger will keep pumping poison from the attached storage sack even after the bee has flown away to die (once having lost its sting, the bee has technically performed its final service to the hive, that of defending it; the stinger does not regenerate, and the bee dies). Stingers stuck in clothing can still inject poison when the skin comes into contact with them, so brush them off as well. *Don't pick off* a stinger, since by doing this you will squeeze the poison sack and, as with a hypodermic needle, inject yourself even more quickly with poison.

That you're going to get stung occasionally is a fact. How often depends on your tact. On the other hand, after a season of keeping bees you will probably have been initiated to that society of semi-immune beekeepers for whom a sting is not much more painful than a mosquito bite and no swelling follows.

REQUEENING

Requeening is very important if you wish to maintain a thriving bee colony. After a long spring and summer laying session the old queen's egg-laying activity tapers off. The colony will prepare for her eventual demise by making sure new queens are forthcoming. The whole process is exceedingly interesting and well worth reading up on. All you need to know as a beginner, however, is that by late fall the hive will be better off with a fresh queen so a large, healthy brood will be ready in time for the spring nectar-gathering. You need a populous hive for good honey production. It is quite possible to let the hive raise its own queen, a process known as supersedure. But importing one ensures getting fresh, good blood into the colony.

Order a queen from a reputable dealer. Instructions will come with the new queen. Check them before introduction to be sure you're familiar with how her particular model of cage works.

Open your hive and locate the old queen. Trap her in an old queen cage—the one you saved from the original delivery—and seal it off so she can't escape. Take her from the hive, but keep her fed in her isolation box. If the new queen is not accepted by the colony, then you have the old one in reserve until yet another one can be shipped.

Approved scheduling calls for removing the old queen in the morning and introducing the new one the next afternoon. Don't wait longer than forty-eight hours, or the bees will start raising a queen of their own.

The new queen can be left in her introduction cage, to give the bees a chance to adjust to a stranger with a strange smell. Remove the attendants that came with her—the queen cannot be shipped alone, because she is so highly specialized that she can't even eat, but has to be fed by other bees. Do so inside your house, just in case she escapes with them. It's easier to find a queen bee on the living room floor than in the hay meadow.

In most cases, a bee colony will accept foreign royalty immediately. But the extra precaution of leaving her in her cage is well worthwhile. Should you introduce a queen directly and the bees not take to her, they would kill her by balling. That is to say, they would just pile up on her in one big suffocating ball. If the bees do not cluster, trying to ball the new queen's cage, everything should be all right. Tear the cardboard off the end of the new queen's introduction cage, exposing the candy, which the bees will gnaw away to liberate the queen.

Close up the requeened hive and leave it alone for five to six days to give the new monarch a chance to settle in. After that remove the cage. As for the old queen—off with her head. You can't just banish her or she'll find her way back to the hive. Squash her.

SWARMING

Swarming has its good points and its bad. In any case, it's almost inevitable, so you might as well make the best of it—and get a free colony out of the deal.

Essentially, swarming is the natural way bee colonies expand. It's a basic part of the total bee cycle. A hive becomes overcrowded, the young Turks, restless, decide to take off with a new queen; they realize the old monarchy is crumbling. This rebel crusading instinct in bees is another reason for requeening annually. With a strong young queen in the hive, swarming is less likely.

How can you tell if your bees are going to swarm? Well, the weather is one clue. And you can't do much about it if it just feels right for swarming. That's one reason why you want to make sure the hive is shaded during the hottest part of the day. Brood chamber reversal also helps discourage swarming. And since you are requeening each year, you don't need the queens being raised by the colony. Inspect the hive once a week for queen cells, almost peanut-shaped and with plenty of extra wax built up into a tip. Unless you want home-grown queens to start additional colonies, destroy any queen cells by crushing them. No extra queens, no swarming.

Still, there are times when no matter how many preventative measures you've taken, your colony will swarm. Or perhaps you will decide to let it swarm so you can start another hive. Most beekeepers build onto their apiary eventually. If you're planning on expanding, order new hives early in the spring so they will be set up and ready any time the bees decide to make their move.

A swarm begins with a large number of bees clustering on the outside of the hive. Suddenly someone inside yells fire and the rest of the hive empties. Somewhere in the center of a rising black cloud of thousands of bees is a queen. Follow the swarm; they're in a great mood and about as calm as bees can get. Usually they won't go too far before settling down, but plan on spending the better part of the day chasing and recapturing them. Where they perch may vary from a convenient spot like a low-hanging tree branch to something more unorthodox like the bumper of your car or the top of the pump. Wherever they settle, shake or brush them into a large box or basket, and carry them back to a waiting new hive. Sticking the box in a burlap bag is helpful if you're a long way from the hive. If the swarm happened to land on a branch, you can sometimes just saw it off and carry the whole thing home as is.

Dump the bees off in front of the hive so that they spill out on the alighting board at the entrance. The bees will crawl all over the place, hopefully some of them into the hive. One thing that helps coax them into their new home is a frame of brood from an established colony. Make sure all the bees from the old hive are removed from the brood chamber before inserting it in a new hive.

If the resettled bees start exiting en masse again, they've probably mislaid their queen somewhere. Look around the outside of the hive and on the ground. Once you find her, give her a helping hand. If she is nowhere to be seen, return to the place the swarm first settled; you'll probably find a small group of bees still clinging to the renegade queen.

Often the queen to leave a hive is the original one. You can order your packaged bee queen with clipped wings. In that case, if she leaves the hive she'll fall flat on her face. The swarm will remain around her on the ground. All you have to do is scoop the queen up and drop her into the new waiting hive. It saves a lot of chasing.

WINTERING

Till recently sugar syrup was used as a honey substitute to tide the bees over the winter and spring till nectar-gathering time. The beekeeper wanted to take the maximum amount of honey from his hives, and a sugar syrup replacement was deemed cheap. Although this method is still practiced in some apiaries, the modern trend is toward leaving enough honey in the hive for the winter and then building up the colony with packaged bees for the spring nectar harvest. The honey-wintering method is ideal for the apprentice farmer, since it minimizes the amount of maintenance work needed on the hive. Unless you're planning to retail massive amounts of honey, you won't even have to buy additional bees in the spring. The colony can build itself up and still produce thirty to seventy pounds of honey a year. Only the supers are removed for honey. The food chamber, as its name implies, then supplies enough honey for a healthy wintering of the whole colony. In areas north of, say, Virginia, a deep super of filled honeycomb should be left at the top of the hive. South of this latitude, a shallow, or half-super will suffice to ensure a good winter store to tide the bees over. No messy sugar syrups or other problems. In spring you put the supers back on and begin your honey harvest all over again.

South of latitude forty, providing a winter food chamber is sufficient preparation for the season. A windbreak may be set up, but no packaging of the hives is required.

Wintering in the northern states, however, is quite another problem. Here a windbreak is absolutely essential. And you've got to bundle up the hives, reserve food chamber and all, if you expect the temperature to go below 20°F. for long periods of time. Loose straw or fiberglass should be laid to a depth of about four to six inches on top of the hive. The sides should have no more than two inches of straw insulation or one inch of fiberglass. Too much insulation would keep the cold in, not letting the hive warm up during the day. For the exact amount of insulation needed in your area, ask the advice of a local apiarist if there's one nearby.

The straw- or glass-insulated hive should be covered with a layer of thin tar paper. Don't use plastic; it can't breathe. The tar paper should be tied on firmly with string. But keep it loose, leaving air holes at the top four corners to permit moist air to escape, and an opening at the bottom entrance. A top entrance should be made by drilling a three-quarter-inch hole through the side of the food chamber at a level with the hand holds. The food-chamber entrance is important for wintering, since the ground-level one often becomes clogged with dead bees or ice. Also it permits better ventilation, another important factor in winter survival.

Once you've packed your hive for winter, let it be. Opening it to see how the hive is doing is a good way to start a dysentery epidemic. In addition, cold bees will consume extra food, cutting into the supply.

DISEASES

Wax Moth. The greater wax moth, *Galleria mellonella,* lays eggs from which emerge grayish-white larvae that like to burrow into wax. They destroy the combs with their tunneling and munching, and build cocoons on the frames. Once they have attacked a weak colony, it's all over. Strong colonies, however, hustle the parents out the door and kill them before they have a chance to lay their eggs. Keep your colony strong, and your wax moth problem will be nonexistent. All spare wax, such as foundation or combs, should be stored in airtight containers with moth crystals suspended from the top in a cloth bag. Of course, this will mean airing the wax well after storage before it is used in the hive. If the wax moth does manage to invade a hive, the only way to eliminate the larvae is with hydrocyanic acid gas or methyl bromide. Both are deadly poison to all living animals and have no place on a natural farm.

Foulbroods. The two most destructive enemies of an apiary are the foulbroods, American (AFB) and European (EFB). The "foul" refers to the odor, not the fouling up of the apiary—but that's what it does. The smell of AFB in its early stages is that of rancid glue, in the latter stages that of a two-months-dead sheep. The foulbroods attack the bee larvae within their growing cells, killing them. Your first couple of hive inspections will have shown you what healthy larvae sealed in their cells or on their way to being sealed look like. Foulbrood-affected larvae, on the other hand, will be yellowish-brown in the early stages of the attack, and "melted" into globs with the progression of the disease. Call the bee inspector. The odor, usually the first hint of trouble, comes from the decaying larvae. The normally fastidious bees let them lie and rot. EFB attacks the larvae in an earlier stage of development; there's less to rot, so the smell isn't so bad.

But it's not the smell that's the trouble, it's the loss of colonies. In the case of verified AFB, the only solution is to seal off the affected hive at midnight when all the bees are in. Cover it with plastic to suffocate them, then burn the whole lot, bees, hive and all, in a blazing pit at least two feet deep. Bury the ashes.

The bee inspector will do the burning for you, and although he will let you get away with burning only the bees, frames, and combs, you're better off taking the economic loss of destroying the whole hive. It could be saved by scorching the inside with blazing gasoline. But safety pays where foulbrood is concerned. Being highly contagious—the dormant spores can attack after fifty years—it could wipe out American beekeeping if it ever got out of control.

Destruction of the country's apiaries would mean a mass reduction in crops, since a vast number of the five thousand or so original species of wild bees that used to do the pollination in this country have been wiped out with pesticides and ecological changes. EFB is not as dangerous, and can be handled without destroying the whole hive. But let the bee inspector decide, because it is sometimes difficult for the apprentice to distinguish between the two varieties of foulbrood. In either case, place future hives as far away from the previous site of the infected hive as possible.

The best preventative against the foulbroods is keeping your colony strong. This means requeening in the fall and supplying an ample food store for the winter. Buying only disease-resistant bees, from a reputable dealer, and regular bee inspection by the local agent are also essential.

Dysentery. Dysentery in bees is caused by spoiled or insufficient winter stores and inactivity brought about by a severe, prolonged winter. The diagnostic hallmark of the disease is numerous streaks and spots along the sides of the hive, caused by constant diarrhea. In severe cases, the colony may be doomed. If the spring is warm, with plenty of sunshine, they'll survive, but in a weakened condition. Should they appear to be improving by April, order two to three pounds of packaged bees to help strengthen the colony. Otherwise it will be easy prey for other problems.

If you see a few yellow spots on the hive in spring, dysentery is probably not the culprit. Even healthy bees will spot to some extent with the change in diet that comes with spring. Again, maintaining a strong, well-wintered colony is the best protection you can give the hive.

If it's beginning to sound like the one effective remedy for almost all of your bee ailments is maintenance of a good colony—it is. This is what makes beekeeping so easy. Careful maintenance solves almost all problems before they can start.

HONEY

As interesting as beekeeping is—and you'll no doubt spend many hours watching the bees at work—what you're after is the honey. As an aside, lest you feel guilty about watching the bees slave away while not working yourself, the old phrase "busy as a bee" is only partially true. For all their activity, individual bees are not that overworked except briefly during the main honey flow. Research has shown that the average bee spends half his time loafing around the hive doing nothing constructive whatsoever.

The two popular forms of honey are extracted and comb honey. Extracted honey is drained from the comb with a centrifugal extractor. To get comb honey, each frame in a super is subdivided by mini-frames about four and a half inches square so the bees will make individual chunks of honey each slightly less than a pound in weight. Comb honey is relatively the more difficult to produce of the two. Although you'll no doubt want to try it eventually, in the beginning plan on extract honey only.

You should have at least two, and preferably three, supers. Once the filled cells in the first one are beginning to be capped over, put a second one on top of it. When the first one is at least three-fourths capped over, pull it out and put it on top of the second, emptier one with a bee escape between them. It's the same type of reversal you performed on the brood chambers. The bees will then go down into the emptier of the two supers and the main hive. They will not, however, because of the bee escape, be able to return. If you have supers enough, you can leave all of them on till fall for convenient one-time extracting. Just make sure they're at least three-quarters capped over, or the honey will turn sour. When using more than two supers, the empty one is always introduced on top of the very bottom one that is still being filled. Any fully filled supers sit above these two.

Two days after a bee escape has been inserted below it, a super filled with honey will be empty of bees. You can now take it off the hive and extract your honey. If the honey flow is very heavy, three, four, five, or even six supers may be used. Since a super loaded with honey weighs sixty to seventy pounds, the amount of work involved is considerable. However, half-supers, or shallow supers, hold only about half as much. Use them if possible. They'll have to be replaced more frequently, since they fill up quickly, if you extract your honey one super at a time. Or simply get extra half-supers and leave that many more on the hive. Shallow supers are much easier to work with. You'll handle them more quickly, and the actual elapsed time needed to empty two of them is not much greater than that for one full-depth super. If you're new on the farm, previously having spent many years at less strenuous work, it will save your back.

Honey is held in the comb cells by capillary action. That's why it doesn't run out as fast as the bee puts it in, and why even uncapped honey remains in its cell. This also is why you can't simply turn the combs on their sides and expect to get honey pouring out. You need an extractor that will spin it out with centrifugal force.

There are various models of extractors, ranging from two- to fifty-frame capacity, and hand-cranked to motor-cranked. The most suitable one for the small apiary is a basic three-frame extractor, with removable frame-holder so that the caps can also be spun free of honey in a separate operation using the same machine.

You'll need a screened shed or porch in which to work. Bees, of course, love honey, and if you extract it in the open, you'll soon find yourself surrounded by your colony bent on getting their crop back. The extractor should be bolted to a box or low table so the honey can flow directly into a bucket placed beneath the spigot. Preferably, the work area should have a concrete or other washable floor that can be hosed down. Walking on spilled sugar is a very noisy and annoying proposition. Walking on spilled honey is even tackier.

In addition to the extractor and a place to put it, you should have ready two uncapping knives, a tin can you can fill with boiling water to keep the knives warm, a galvanized wash pan or tub (a store-bought uncapping can is more convenient but not necessary), and several square five-gallon honey tins for storage. A five-gallon tin will hold sixty pounds of honey. You can expect this much from each colony— more if the flow was heavy that season. For pure honey, without any suspended pollen or bits of comb and other scraps, you'll also need a frame over which you can stretch dampened cheesecloth as a filter. Better yet is a large, long funnel lined with cheesecloth. By using a funnel and keeping the end of the tube always submerged in the honey, you'll avoid most of the air bubbles that the frame-filter method puts in the honey.

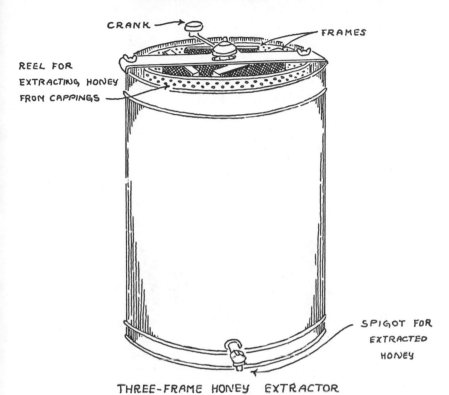

CRANK

FRAMES

REEL FOR
EXTRACTING HONEY
FROM CAPPINGS

SPIGOT FOR
EXTRACTED
HONEY

THREE-FRAME HONEY EXTRACTOR

You don't have to extract the honey immediately upon taking a super off the hive. However, it's the preferred way. At this stage the honey is still warm and thin, making your job easier. Place the super near the extractor. Take out a frame. Put it on the washtub. For easier working, you may want to lay a board notched to fit the tub across it about a third of the way from its center. A nail hammered point up through the board will give you a convenient pivot on which to rest the end of the frame.

Take a warmed uncapping knife from the can of hot water. Holding the frame so it leans over the tub, draw the knife slowly upward, cutting the caps off. The angle of the frame should be such that the cappings fall clear of the knife and into the tub. When you've finished one side of the frame, spin it around on the nail and uncap the other side. Put the knife back in the hot water and stick the frame in the extractor. Once you've uncapped the number of frames your machine will hold, you're ready to extract.

Turn the reel of your extractor slowly at first. You'll notice that

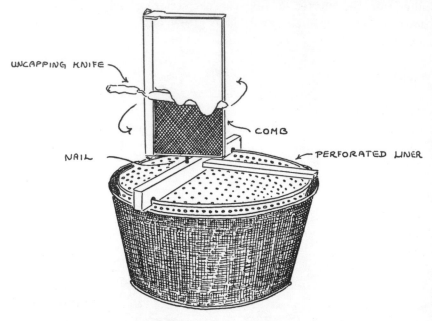

UNCAPPING KNIFE

COMB

NAIL

PERFORATED LINER

UNCAPPING TUB

CUT THE CAPS, OR COVERS, OFF HONEYCOMB WITH A SMOOTH UPWARD MOVEMENT OF A WARM KNIFE. CAPS WILL FALL INTO PERFORATED INNER LINER. THE HONEY CLINGING TO THE CAPS WILL DRIP INTO THE TUB. AFTER UNCAPPING ONE SIDE OF THE COMB, SPIN IT AROUND AND UNCAP THE SECOND SIDE.

the action of the machine is not entirely smooth. This is because not all the frames weigh exactly the same, causing imbalance. And it's the reason the machine must be bolted down. As soon as it is turning more smoothly, speed up, until about half the honey is extracted from each comb on the side facing out. Now reverse the frames in the extractor. You want speed, but be careful you don't go so fast that the combs break from the centrifugal force. You won't know how fast this is until you've actually broken one, of course. It's no major loss, but it means putting a new wax foundation in the frame, whereas in the case of well-extracted frames, you have only to return them to the hive for a refill. The second side of each comb should be spun clean, before reversing the frames a final time. Now you spin the first side, still half full of honey, entirely clean.

The honey will run from the extractor through your cheesecloth filter into the bucket or tin. The cheesecloth should be wet before starting, and washed out occasionally, depending on how much extraneous matter is in your honey. If you intend to sell the honey, it must be filtered this way to give a more pleasing appearance. For your own use, however, consider passing up the filtering step. Your honey will have a little wax in it, but that's not harmful to the digestion. As to the pollen, it is a rich natural food filled with proteins, minerals, oils, and sugar. Its removal is only part of modern man's purification fixation. There will also be, of course, an occasional bee leg or such, but certainly nothing to cause you any harm.

Once all the frames have been emptied, return them to the hive. The bees will now oblige by cleaning up any remaining honey bits and refilling the combs. When you're finished extracting from the combs, you'll still have a tub full of cappings and honey. Remove the frameholder from your extractor and pour the contents of the tub evenly around the extractor's wire cage. Spin it and you will get more honey. Stir up the cappings and spin again.

The caps themselves can be saved and used in making beeswax candles. Because of the relatively low melting point of beeswax, however, remember that these candles burn fairly quickly, so plan on fat ones.

Clean all your extracting equipment thoroughly with tepid water when you're through. Then sit down in that proverbial porch rocker for some hot homemade rolls with butter and fresh honey. It's a rewarding way to watch the sunset.

The Larder

*Let them make sausage of me and
serve me up to the students.*
—ARISTOPHANES

O NE OF MY FONDEST MEMORIES of childhood was sneaking into the ice shed and chipping off a sawdust-flavored "popsicle" in the heat of midsummer. The ice shed wasn't a house where an ice machine was kept—our ice machine was a pond in winter. The ice was cut into large blocks from the center of the lake by timber saw and then hauled by sled to the ice shed, where it was covered with layers of insulating sawdust. Come summer, the frozen pond still fed the icebox. A real icebox.

In winter the memories were made from the earthy smell of the root cellar, from the salted meat and fish, the onions and the dried rose hips in the pantry, and from the kitchen wood box. Often I wonder how the world has been able to so thoroughly obliterate these glorious aromas. And then I find, in the country again, that it hasn't.

ROOT CELLARS

For winter storage without refrigeration, there are three distinct conditions needed by various crops: cool and moist, cool and dry, and warm and dry.

Often the heating unit for the house keeps the modern basement

304

too warm and dry for use as a cold cellar. This, however, makes it ideal for crops such as the pumpkins and squashes, which store best in precisely such surroundings. Use only the healthiest specimens for storage. Put them on shelves that permit air circulation, but not something like metal rods or wires that would cut into the fruit. Leave room between the pumpkins or squashes so the air can get between them. A temperature in the vicinity of 50°F. is good.

Building a root cellar is well worth your while for both cool-moist and medium-moist storage. The area toward the bottom of the cellar will be moistest, that toward the top driest. The root crops—beets (leave an inch of green when storing or they will bleed), carrots, rutabagas, turnips, and the like—should get the moistest areas. Potatoes, cabbage, and cauliflower need the slightly drier area, as do your apples

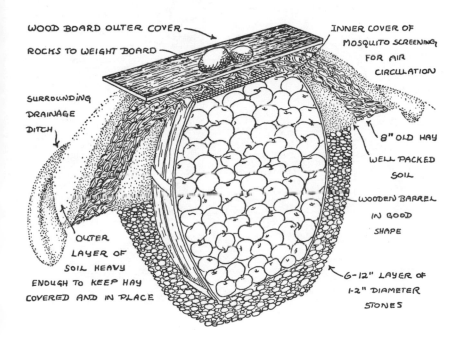

WOOD BOARD OUTER COVER

ROCKS TO WEIGHT BOARD

INNER COVER OF MOSQUITO SCREENING FOR AIR CIRCULATION

SURROUNDING DRAINAGE DITCH

8" OLD HAY

WELL PACKED SOIL

WOODEN BARREL IN GOOD SHAPE

OUTER LAYER OF SOIL HEAVY ENOUGH TO KEEP HAY COVERED AND IN PLACE

6-12" LAYER OF 1-2" DIAMETER STONES

FRUIT STORAGE BARREL

REMEMBER ONE ROTTEN APPLE SPOILS THE BARREL. STORE ONLY FRUIT THAT IS NOT BRUISED AND HAS NO WORM HOLES OR OTHER FLAWS. WHEN TEMPERATURE FALLS BELOW 25°F, REPLACE INNER SCREEN COVER WITH A SOLID WOOD ONE.

and pears. A temperature in the range of just above 32°F. up to 40°F. is what you want for them all. To this end, insulation is necessary. Just burying the root cellar in the earth, mounding it over with the excavated dirt, is sufficient in most areas. Plant the roof to a cover crop and crocuses. In any case, you don't want the root cellar sealed off in any way. Ventilation is essential.

If possible, the fruits and the vegetables should be stored in two different cellars, since the fruits have a tendency to absorb odors. At least be sure to keep them as far apart as you can, if you haven't actually got two cellars yet. Dividing a root cellar into two rooms separated by a door works well.

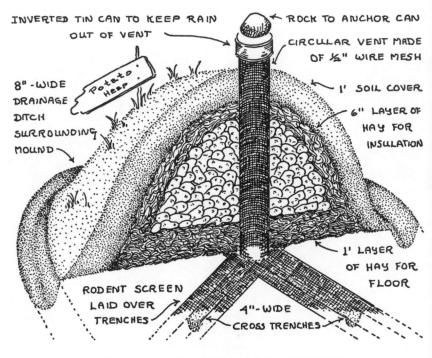

INVERTED TIN CAN TO KEEP RAIN OUT OF VENT

ROCK TO ANCHOR CAN

CIRCULAR VENT MADE OF ½" WIRE MESH

8"-WIDE DRAINAGE DITCH SURROUNDING MOUND

Potato Heap

1' SOIL COVER

6" LAYER OF HAY FOR INSULATION

1' LAYER OF HAY FOR FLOOR

RODENT SCREEN LAID OVER TRENCHES

4"-WIDE CROSS TRENCHES

ROOT CROP STORAGE MOUND

MAKE THE STORAGE MOUND HALF AS HIGH AS IT IS WIDE. ALLOW ABOUT A FOOT OF DIAMETER PER PERSON. MEASUREMENTS GIVEN ARE APPROXIMATE. THIS NEED NOT BE A PRECISION JOB. WHEN THE TEMPERATURE FALLS BELOW 25°F, CLOSE OFF ENDS OF TRENCHES WITH PACKED SOIL AND PUT A BIGGER CAN ON TOP SO THAT ONLY ½" OF SCREENING IS LEFT UNCOVERED.

You'll have enough work to do on the farm the first year without building a full-fledged root cellar. But that doesn't preclude storage. Temporary root cellars for individual crops have long been in use. A half-buried barrel of apples will carry you through till spring. A cabbage mound will handle that crop. Another mound for potatoes and a third one for mixed root crops, and you're all set for the winter.

Onions, garlic, and beans need dry, cool storage. Hang them inside a shed unoccupied by livestock. An occupied barn would be too warm and moist. Your livestock would probably get to them before you could, so they wouldn't necessarily be wasted, but that's not what you're hanging them for.

Lest you be disappointed after eating garden-fresh vegetables all summer by the flavor of your storage fruits and vegetables, one might as well admit right here and now that they'll be about on a par with those you've been buying at the supermarket. Also, you're not going to have apples all year round, nor strawberries nor grapes. But the deprivation will just make you appreciate them more. Today's lifestyle of always being able to buy what you want, if perhaps not of the best quality, has leveled life to a kind of seasonless mediocrity. You'll be amazed how wonderful an apple tastes when you've grown it yourself, waited for it, when your mouth began to water with the first apple blossom and continued to water as the fruit slowly formed. That first bite of a fresh, ripe apple then tastes every bit as good as it did to Adam.

KRAUTING

Salting is one of the oldest methods of preserving food. In the northern European countries, where long winters and short growing seasons made food preservation a major key to survival, it was, throughout the Middle Ages and the Renaissance, almost the exclusive means of hoarding a winter larder. Unfortunately, a winter diet of solely salted and pickled meats, fish, and vegetables is a poor diet; and the health of those dependent on this form of nourishment was often severely impaired. But that doesn't mean the method has to be discarded entirely. Today's winter larder can be well stocked with home-canned and fresh-frozen food. And wintering livestock has become much more economical. Salting is used in the modern country kitchen more to give pleasant variety than to really supplement the regular diet. The family kraut barrel is one tradition that's certainly worth resurrecting—provided, of course, that you like sauerkraut. The bean crock is another, by way of variation on the theme.

THE KRAUT BARREL

You'll need a large container, either an old sixteen-quart or bigger earthenware crock or a barrel. An old pickling barrel is made to order for krauting, and some nail kegs are usable—but make sure when using wood that it's hardwood, not pine. Pine-flavored kraut does not rank high on the epicure's scale. Also, a wooden barrel that has not been used previously to store liquids will have to be soaked in water for about a week to swell it. Whiskey barrels, which are used only once commercially, are ideal if you can find them.

Whatever you use, it must be "square" when it is filled; that is, the kraut should occupy an area as deep as it is wide. Be it a barrel, or a crock—which I personally prefer—or glass jugs, the container should be thoroughly scrubbed and scalded before use. Sun-drying is best.

THE CABBAGE

Trim off all the wilted outside leaves from a batch of freshly picked cabbages. Add the waste to your compost pile or, if there's not too much of it, feed it to the pigs. (Too much cabbage will bloat grain-fed hogs.) Wash all the heads of cabbage, cut them in half, and start shredding. Try to get hold of a traditional kraut cutter for this. It's nothing but a flat wooden board with a sharp slanted blade affixed over a slot in the center through which the sliced cabbage falls, but it makes for a lot easier and even-tempered shredding. The only problem is, a good kraut cutter is hard to come by in this day and age. You may have to settle for some other type of vegetable slicer. Whatever the case, don't count on doing it with a knife. Consistent thinness of the shreds (about one sixteenth to one eighth of an inch) is the key to well-cured sauerkraut. Also, watch your fingertips when using a shredding board—it's not the best way to trim your nails. Things move quickly when you've got twenty pounds of cabbage to slice while sitting around the kitchen table chatting. And you'll need twenty pounds to make eight or ten quarts of the finished product.

When you've got a decent tubful of cabbage shavings, roll up your sleeves and mix the cabbage around with your hands so that any pieces clinging together are broken up. But don't bruise them.

THE SALT

You'll need salt at the rate of one cup for each twenty pounds of cabbage. "Free-flowing" salt can't be used because it contains chemicals that kill the necessary bacteria. You need the old kind of salt that cakes up when it's humid. Either granulated or flake salt will do. Sea salt is good, but if you get it in hunks, you'll have to grind it before using. Very coarse salt won't do the job. The salt for a kraut barrel is to aid fermentation. You're not using it as a preservative agent, as you would be in making, say, salt pork, where a much greater quantity is added.

Working in small batches, say, five pounds of cabbage at a time, sprinkle in the salt slowly, in this case a fourth of a cup, while tossing the cabbage shreds gently in a bowl. Place the well-mixed slaw in your barrel or crock in two-inch layers, pressing each layer down with a clean chunk of two-by-four, an old zigzag potato masher, or whatever similar blunt object is handy. Again, remember not to bruise. You want to compact the cabbage so it will be covered by its own salty brine in the end. You don't want to mash it. While you're layering in the cabbage, you can also sprinkle in some caraway seeds for extra flavor— no more than about a teaspoon per quart unless you're as crazy about caraway as I am, for it can be strong stuff.

BRINGING UP THE BRINE

The ubiquitous cheesecloth enters the picture. Cut six pieces each large enough to cover your jug or barrel and then some for the edge. Once the crock is filled to within two inches of the rim, cover the well-tamped cabbage with the layers of cheesecloth, tucking the extra inch or so down between the slaw and the wall of the crock.

You will need a follower. This is simply a disc just a shade smaller than the inside diameter of your container placed on top of the kraut to help keep pressure evenly distributed over the mix. Don't use a metal disc (metal will react with the acid), pine (lousy flavor), or plywood (same reason). A hardwood disc is best. If you can find a snug-fitting plate, it makes a reasonable substitute placed curved side down as long as no air is trapped between it and the cabbage. The follower is there not only to help squeeze enough juice from the kraut to cover it with brine, but also to keep air out.

To insure an airless brewery, a weight should be placed on top of the disc or plate. A gallon jug is convenient if it will fit atop the follower; it can be filled with just enough water to produce the desired pressure. A big fat stone also works fine, and is more like the real thing. Six hours after the weight has been applied, the brine should have risen sufficiently to cover the top layer of cabbage. If it hasn't, add more water to the jug or more big fat stones.

FERMENTATION

Keep your weighted kraut barrel in a well-ventilated spot where the temperature stays between 65° and 70°F. Much above this range and you'll have what is politely called slimy kraut, which means it's spoiled. Much below, and fermentation will take forever.

The actual fermentation process produces gas. Peeking under the lid of the barrel, you'll notice small bubbles rising to the surface. You'll also smell them. When there are no more bubbles, usually in around four weeks, your homemade sauerkraut is done. But meanwhile, during the fermentation process, a white scum will keep forming at the top of your mix. This must be removed regularly, usually daily, although it's permissible to skim it off only every second day. Simply remove the weight and follower, then lift the cheesecloth gently so the scum will cling to it. Rinse the cheesecloth well in very hot water. Don't use soap. But cool it off again under cold water before replacing it. A steamy blanket on top of the kraut encourages spoilage.

If during fermentation the brine level in the barrel sinks below the top layer of kraut, fresh liquid must be added to top it up. A couple of pinches of salt in a quart of water will usually be enough to replace the lost brine. Technically, plain water would do, since no salt is lost in evaporation, but the cheesecloth may have skimmed off a bit of salt lick in its daily trip to the sink.

STORING

Once the bubbling in the barrel has stopped, usually in about four weeks, remove the weight and follower permanently. The sauerkraut is now ready for eating. If it's your first batch, you'll probably want to sample it right away. However, ten quarts of sauerkraut is a lot to put away at one sitting. Besides, the reason for making it in the first place was to store it against the wintertime when your garden wouldn't be supplying you with fresh vegetables.

To store kraut, either in its original crock or in sterilized smaller jars, all you need to do is make certain it's still covered to the top with brine, add a layer of paraffin or beeswax to seal out all the air, and put it in a cold place, below 50°F. but above freezing.

The acidic property of well-fermented kraut, combined with the lack of air when well sealed, eliminates the growth of spoilage bacteria and mold, provided the sauerkraut is kept in cold enough storage. If it has a dubious look about it when you open up a jar—appearing pinkish, slimy, rotten, or black—there was probably something wrong with the way it was prepared. Maybe it was exposed to air, or over- or undersalted, or the scum not properly removed. If so, add that batch of kraut to your compost heap and start again. However, with careful preparation you should have no problems. Making sauerkraut is a basic, easy process, and a well-fermented, well-sealed batch will keep safely for three to four months or longer.

So will string beans—pardon me, snap beans. Just french them first and prepare the same way you did the cabbage, except don't use caraway. Makes a tasty change from kraut.

DRYING FRUIT

Many fruits—peaches, plums, or cherries, for instance—do not store well in the cold cellar for more than a few weeks because they are soft and easily bruised. This, however, doesn't mean they can't be kept—only that the method of preservation is different.

Dried fruit has been an important dietary staple for centuries. But the commercially dried fruit available today is riddled with sulfur. The function of sulfur is essentially that of a cosmetic: it keeps the fruit from darkening in color. In some cases, it also tenderizes it. Sulfur in small quantities may be of some use as a trace element in the diet. The quantities consumed when dried fruits are made a steady dietary supplement, however, certainly can't be beneficial. Fruit does not have to be sulfured for storage. Naturally dried fruits will be darker and a little chewier—but they'll be all fruit.

Most fruit can be dried. Don't use bruised ones or those with any signs of mold, unusual dark spots, or other defects. Use your good fruit. Remember the old saying, one rotten apple spoils the barrel? It does.

First wash the fruit and dry it with a towel. Then cut the fruit into pieces, removing pits, stones, or other inedible portions. The smaller the pieces, the quicker the drying. Juicier fruits need smaller pieces. Apples and pears may be quartered; peaches, the same, since the fruit is small. Do your cutting over a bowl and you'll have plenty of nectar when your work is done.

Sun-dried fruit is the best, if you have dry autumns. Place the cut fruit one layer deep in a double-screen drying tray, screen-topped to

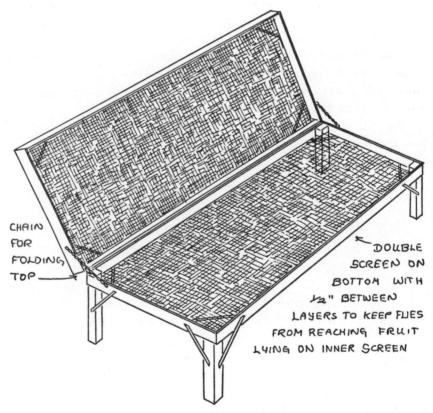

CHAIN FOR FOLDING TOP

DOUBLE SCREEN ON BOTTOM WITH ½" BETWEEN LAYERS TO KEEP FLIES FROM REACHING FRUIT LYING ON INNER SCREEN

FRUIT DRYING RACK

MAKE TO SIZE CONVENIENT FOR YOU FROM SCRAP LUMBER.

keep out the bugs. Keep each piece from touching the next, to increase air circulation. Turn the fruit two or three times a day. Take the tray indoors at night, or the dew will undo the whole day's drying. The fruit is ready for storage in glass jars when the outside is quite dry but a piece is still soft enough to bend with your fingers. You don't want it brittle, but you shouldn't be able to squeeze any juice out of it either.

If you have wet and rainy autumns, oven-dry your fruit. Set the temperature at 130°F. Again, screen-bottomed trays are best, but of course they don't have to have the bug lid. Average oven-drying time is four to six hours.

MAKING CHEESE

Most of what you grow in a time of plenty can be stored as surplus for future use in bleaker days. But what about milk? Refrigeration will hold it for two weeks, but not longer effectively. And the ancient tiller of the soil had no refrigeration. His milkers were dry many more months of the year than today; yet milk, by necessity, was a major part of his diet. Somewhere along the chain of agricultural development, probably not too far after the discovery of yogurt, came cheese. Solidified milk that not only lasted months but improved in flavor as the days passed.

There are countless varieties of cheese in the world, from soft to hard, mild to sharp, white to blue or brown. The French alone make three hundred kinds, although most are simply regional variations on a theme. Almost all of them are produced in basically the same way, their distinctive quality being the outcome of bacterial conditions, means of storage, additives, aging, and, of course, type of milk used— skim or whole; from cows, ewes, goats, buffalo, llamas, or even yaks.

For the beginner with a small herd of goats, a variant of "American" whole-milk cheese is perhaps the best. It lends itself well to production from goat's milk, takes no expensive equipment, and is surprisingly easy to make. The one thing to remember in planning a cheese cellar is that the volume of your yield shrinks considerably as you progress from milk to cheese. Don't think that by starting with five gallons of milk you'll end up with a year's supply of cheese. You're going to nibble up this delicious homemade dairy product faster than you think.

Cheese is made by coagulating milk, removing the whey, or liquid, and preserving the curd, or solid part. Coagulation is achieved through the addition of rennet, a natural salt-brine extraction from the fourth stomach of young milk-fed calves (cows have four stomachs). Rennin is an enzyme that acts upon casein, the chief protein found in milk. Although the chemistry is complex, you really don't have to know all about it to get started. The fact that you can buy rennet readily at the country store is enough. There'll be time for more reading on the subject some long winter's evening when the harvest is in and the chickens are lying low—over a hunk of your own sausage and a wedge of your first wheel of cheese. Since cheese is a natural product, as long as you follow the tried and true methods you'll have good results.

THE MAKINGS FOR A CHEESE

The Milk. For a family-sized wheel of cheese, start with five gallons of milk (twenty quarts). If, as is likely, your goat herd is too small to supply this much milk in a day, it's quite all right to keep the milk in cool storage (50° to 60°F.) till you have enough. A week is about the practical limit. Be sure it's in well-sealed containers so it doesn't pick up any odors, which would spoil your cheese before you even start.

The Starter. If you're in dairy country, commercial starter containing *Streptococcus lactis* can be bought locally. However, a satisfactory home-kitchen "starter substitute" for five gallons of milk is one cup of cultured buttermilk. In the distant past cheese was always made from milk in the process of souring. Today, for more consistent results, fresh milk is used with the necessary bacteria being introduced as the starter.

Stir the starter into the milk and keep stirring while the mixture is heated slowly to a temperature of 85°F. You have a little flexibility with the temperature, but try to stay as close to 85°F. as possible. Use of a double boiler is recommended to keep the milk from scorching. Putting the milk pot in a large tub of water over the stove will do admirably.

Rennet. Once the desired temperature of 84° to 86°F. is reached, you're ready to add the rennet. Use one full teaspoon of rennet freshly mixed with three-fourths of a cup of cold water. Stir the rennet solution into the heated milk slowly.

At this stage food coloring may be added. However, since most of these are chemical in nature, they add an unnecessary artificiality to your cheese. There's nothing wrong with white cheese. Of course, once you've made a few wheels, you'll want to experiment with natural coloring agents such as saffron, achotti, or even licorice extract. Or try some different herbs and spices for piquancy. But don't be surprised at the strange flavors, colors, and failures you may achieve. For your first wheel, stick with a white cheese.

The milk and rennet mixture should be stirred with a wooden cheese paddle, perforated wooden spatula, or just a plain, large flat wooden spoon for one minute, no more than two (count from 101 to 200 slowly). Then let it sit and watch it coagulate. This will take roughly half an hour. At the end of this period you have a firm jelly-like curd. But suspended throughout the curd is whey. The whey must be extracted by cutting before your cheese can go any further.

CUTTING THE CURD

A good way to judge when the curd is ready for cutting is to press the cheese paddle, lightly, flat on the top of the curd near the side of the pan. When ready, the curd should break away clean from where it touches the pot. Insert a knife, one long enough to touch the bottom of the pot, all the way down at the edge, and slice across the whole curd. Repeat the process until you have a series of parallel cuts half an inch apart. Cut these parallel strips into squares by slicing cross-wise, again with half-inch spacing. Now take a wire cheese cutter, or use an angel-food cake breaker, insert it at the edge of the pot, and slice the curd horizontally, all the way to the bottom. After cutting, let the curd sit for fifteen minutes. Whey will drain out while it sits.

HEATING THE CURD

After its fifteen-minute rest, heat the curd *slowly*, stirring constantly with the cheese paddle, in the double boiler or tub till it reaches 100°F. This should take about twenty minutes over low heat. While it's heating, make sure you stir constantly with the cheese

CHEESE   CURD   CUTTER

PULL CUTTER SLOWLY THROUGH
CURD WITH WIRES PARALLEL
TO BOTTOM OF CONTAINER.

paddle. Keep your stirring gentle at the beginning. Be careful not to let the curd scorch on the bottom. After about ten minutes you'll notice the curd stiffening. Keep paddling, but you no longer have to be so gentle.

Once 100°F. is reached, remove the pot from the stove. Stir it up every ten minutes or so. You'll see that as it cools the curd contracts, forcing out more and more whey. Check the temperature occasionally. When it has dropped back to 85°F., keep it at that level; if necessary move the pot to sit nearer the stove again. After forty to sixty minutes of cooling, test a piece of the curd by pressing it between thumb and forefinger. It should feel springy. Release your fingers, and the curd should crumble somewhat in your hand. You're ready to remove the whey.

SEPARATING THE CURD AND WHEY

Letting the curd settle to the bottom for a few minutes and dipping off most of the whey will make your batch easier to handle. Pour off the rest of the whey through a strainer, meanwhile stirring the curd. Otherwise it will stick together and form one lump—which is what you want eventually, but not yet. Your stirring will liberate more and more whey. Drain it off.

Keep a sharp eye on the curd's consistency. This is the trickiest stage of cheese-making. If all the whey is extracted, you'll have a cheese with the consistency of an old Chianti cork. If not enough is removed, you'll end up with a weak, droopy cheese similar in texture to watered-down library paste. Sample the curd as you work. When it feels springy and rubbery, bite into it. If it squeaks as you chew, it's ready. Pour off all the remaining whey and get set to salt the cheese.

SALTING

A little at a time, add one level tablespoon of salt per gallon of curd. If you're making the five-gallon beginner's cheese, you want five spoonfuls of salt. Stir after each sprinkling of salt to mix it in evenly. Check the temperature of the curd occasionally; you still don't want it to fall below 85°F.

HOOPING

The Hoop. If possible, get a small wooden cheese hoop from the general store, the local auction, or a friendly neighbor. Since cheese hoops are going the way of the buggy whip, however, finding one may be a problem. Of course, you can make your own wooden cheese hoop. Or, if your carpentry skill hasn't had enough practice yet for this, you can settle for a No. 10 can (the big size used by institutions and restaurants for their steam-shriveled vegetables). Clean the can and cut out the top and bottom. Be sure you leave no rough edges—you're going to have to slide the finished cheese out. Use the lid you just took off as a template to cut two circles from half-inch plywood. These should fit loosely enough to slide up and down in the can. Of course, if you are lucky enough to find a ready-made cheese hoop, you'll probably get the press and follower—the equivalent of your two circles—with it.

Ever wonder where cheesecloth got its name? Well, here's where. Line the hoop or can with a piece cut wide enough to overlap about an inch at the seam and long enough to fold over and cover the bottom. The hoop should be set in a pan to catch the additional whey you are about to extract. Essentially, what you want is a gauze bucket sitting inside the hoop, covering the bottom and with enough left over at the top to fold in once it's filled with curd so the cheese is completely covered by cloth. You don't need too much extra at the top; the cheese will shrink under pressure. Fold the top of the cloth over the outside, or rim, of the can and put a strong rubberband around it to keep the cloth from falling into the hoop along with the curd.

Ladle about a fourth of your curd into the hoop. Press it down with a wooden spoon all around, making certain all air bubbles are pushed out and the curd is all the way down to the bottom and out to the edges of your hoop. Put in another quarter of the curd, continuing to tamp it down, then another quarter, then the last bit. Now fold the cloth over neatly, put the wooden followers on top, and the cheese is ready for pressing.

PRESSING

Place the hoop, still in its pan, into your cheese press for one hour. It's a good idea to have a flexible weight system. A ten- to twelve-pound weight should be just about right. Slide it up and down the arm to vary the weight. Depending on the amount of whey left, you might need a little more or a little less. What you want is sufficient pressure to give the curd one last squeeze, but not enough to force it out at the bottom of the hoop.

DRESSING

Remove the cheese hoop from the press. Unfold the cloth at both top and bottom and pull to eliminate any wrinkles that may have formed inside the hoop during pressing. Now trim off the excess cloth so that when you fold it back over the top and bottom it meets in the center of the wheel, without much overlap. Stick it back in the press and add another two pounds to your weights. Slowly; don't force down suddenly. Let it press for twenty-four hours in a shed or cellar that's kept at a temperature of around 60°F.

CURING

After removing from the press, keep the cheese in the hoop for two additional days. But lay it on its side so that air can enter freely from what used to be top and bottom. The room temperature should be between 50° and 60°F.

After two or three days the cheese can be easily removed from the

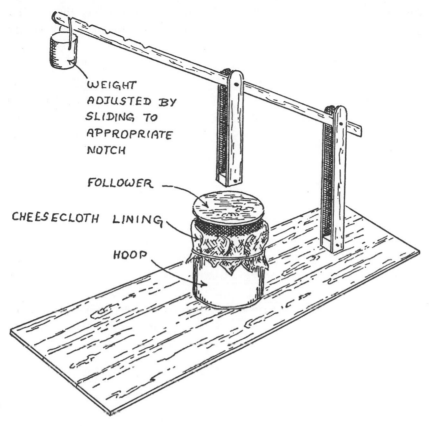

WEIGHT ADJUSTED BY SLIDING TO APPROPRIATE NOTCH

FOLLOWER —

CHEESECLOTH LINING

HOOP

CHEESE PRESS

hoop. Wipe it with a damp, clean cloth, then towel it carefully with a dry one. Bacterial action has helped to make your cheese. Now, however, you don't want any more. To keep mold from forming, the cheese should be wiped clean once a day as it cures, or ages. Keep it on a board, and turn it upside down after each day's cleaning to ensure even drying of both top and bottom. In a week the cheese should have formed a good solid rind.

PARAFFINING

To avoid having to sponge-bathe the cheese every day for as long as you want to keep it, paraffin it. After the rind has formed, melt some paraffin in a double boiler over low heat. Be careful, it burns, and you're not trying for homemade napalm. As it melts, you'll notice it will begin to smoke. From that point on, keep the temperature just high enough so the paraffin stays liquid. Paint on an even layer over the entire cheese with a clean brush. Be sure you cover every spot; Achilles' mother left his heel uncovered and look what happened to him.

Once the cheese has been coated with paraffin, you can store it in your cool cellar for months—if you can resist the temptation to eat it.

CHURNING BUTTER

If you happen to have found a cream separator at a country auction sometime, so much the better when it comes to making butter from your goat's milk. It doesn't separate as readily as that from cows, and a separator will give you something of a head start. Don't let the lack of one stop you, however. Proceed as follows.

Set out the goat's milk in a cool place, using shallow, covered, heat-resistant glass pans. Surface area is important. A quart of milk in a bottle will separate more slowly than the same quantity in several pans. Leave the milk undisturbed for eighteen hours (a little less if the weather's warmish, a little more if it's on the cool side).

After the first setting period, heat the milk very slowly until a light wrinkly skin forms on the surface. Make sure you don't disturb the top cream when you move the containers to and from the stove. Following their heating, the dishes should be allowed to sit for another eighteen hours. Then skim off the cream. This is what you'll be making your butter from. The skim milk can be kept for the lunch milk jug or fed to the pigs as a bonus dessert on their regular menu.

To make butter, you can use a small churn or an electric mixer. Goat's cream makes for *very* slow churning, so don't get discouraged if you're doing it by hand. You may want to blend in just a touch of natural food coloring while you're at it. Goat's butter is white. It's also a little softer and smoother than cow's butter.

As you churn, most of the buttermilk will separate out, while the butter clumps together. Once the clumps are well formed, pour off the buttermilk. (This is real buttermilk, not a waste product.) Then work the new butter with a wooden spoon against the walls of a bowl to squeeze out the last of the buttermilk. Cover the butter with cold water and squeeze it out again. Rinse several times, until the water is clear. If you don't get all the buttermilk out, the butter will keep poorly.

But well-washed butter will keep much longer than you'd expect, even without refrigeration. This is handy for two reasons. First of all, you don't want to put in the work making butter every week. Secondly, it lets you make up extra butter just after the goats have freshened and their milk production is at a peak—provided, of course, you're not busy using all the milk making cheese.

To store butter without refrigeration, mix up a batch of brine, dissolving enough salt in water so that a fresh egg will float. Boil the brine for ten minutes. (Take the egg out first, unless you want hard-boiled eggs.) Let cool overnight. Filter through cheesecloth into a crock. Wrap individual quarter- or half-pound blocks of butter in cheesecloth and submerge them fully. Cover the crock with a weighted dish, making sure the butter is pressed down far enough to be fully submerged. Stored in a cool spot, the butter will keep for six months, so long as you keep the water level up.

Between your crock of butter and your honey pot, not to mention all those berry jams, someone in the family had best start baking up a big supply of home-cracked wheat bread.

MAKING YOGURT

One of the healthiest and simplest of dairy foods to prepare is yogurt. You don't need fancy equipment, although if you're going to make it often, an electric incubator may be worth consideration because it will give you more consistent quality with less watching. But it's a convenience, not a necessity. A thermometer will be sufficient to indicate the desired 100° to 120°F. temperature until you learn to estimate it by dropping some milk on the back of your hand.

The first thing you'll need for yogurt is some starter culture. Get Bulgarian culture. It's the best. Proportions to use and directions come with your starter. The one thing to remember is that the instructions often talk only about cow's milk. Don't let that bother you if you have milk goats. Goat's milk works fine for yogurt.

Yogurt-making depends on bacteria. The right kind of bacteria, however. That's what you're paying for in your culture. So be sure all the utensils you use are very clean before you start.

First bring the quantity of milk specified by your culture instructions to a near-boil. Be sure you don't let it actually reach the boiling point. This would impair flavor. Let cool to the 100° to 120°F. level.

Mix in the culture thoroughly with a wooden spoon, then pour into clean, prewarmed jars or crocks. Give these a warm sitz-bath in water about three-fourths the height of the containers. Keep at the 100° to 120°F. level. When the yogurt is the consistency of heavy cream, it's ready for the refrigerator. If you don't have a refrigerator, plunge the containers into cold water to chill quickly, then store in the coolest spot you've got. It will keep several days without refrigeration.

Save two or three tablespoons of the yogurt as a starter for your next batch. Eventually, however, your home-tended starter culture will weaken. When you no longer get satisfactory yogurt, get some fresh starter.

THE CREAM CHEESE BAG

This one's simpler than unwrapping the tinfoil around those "vegetable-gum-and-fillers" cheese cubes you find in the supermarket. Just mix up a super-large batch of yogurt. Don't make it too tart. Keep what you'll want of the yogurt for lunches, take the rest and pour it into a cloth bag. Several layers of the ubiquitous cheesecloth will do. Hang the bag over a pan to catch the whey. This you can use later for baking or in soups. By the time the yogurt has hung overnight (ten to twelve hours), it's cream cheese. Chop up some chives from the garden and spread on home-baked whole-wheat bread. Makes a meal in itself, it's so nutritious and tasty.

SETTING CLABBERED MILK

The way we used to make "yogurt" was simply to take a couple of spoons of sour milk and add them to a quart of regular milk. Mix and pour out in shallow dishes or casseroles. Let stand on the windowsill, but not in the sun. Usually sets in a day.

Don't try making clabbered milk when there's a thunderstorm around. There's an old wives' tale that says it separates and you get a lot of water around the sides of the bowl when lightning strikes. Old wives' tale or not, ours always did just that. Serve chilled, sprinkled with nutmeg.

SMOKE CURING

The problem with slaughtering a pig is, how do you eat it all? A couple of fifteen-pound hams are a lot to polish off in a week, not to mention the bacon and the pork chops. But you don't have to. Smoke-curing with salt not only lets you keep it for four to six months, but improves the flavor as well. Actually, it's the salt that does the preserving. The smoking aids and abets the process and—its real contribution—adds that sugar-and-hickory touch.

You'll need a smokehouse, unless you've asked the butcher to do your curing for you. The shed is simple to build and can vary in size and shape as long as it meets certain basic requirements. It must—surprise?—be smoke-tight and allow for a constant temperature of between 90° and 100°F. Don't let the temperature go higher, or you'll end up with cooked meat, which will not keep. For this reason the fire pit is usually built to one side of the smokehouse, and the smoke piped in.

Before smoking, the meat must be soaked in brine. The soaking solution contains, besides salt, a little sugar to counteract some of the dryness that salted meat would otherwise have, and saltpeter, which keeps the meat red. Saltpeter is not necessary as long as you're prepared to cut open a nice gray ham for dinner. Remember, by the way, to cook all your pork very, very thoroughly to guard against the possibility of trichina, whichever color you make your hams. Smoking cures the ham for storage; it does *not* cook it.

For a salting brine, mix together:

    1 gallon water
    2 cups salt
    1 cup dark-brown sugar
    ½–1 teaspoon saltpeter

Stir the brine well so that all the ingredients are dissolved before using. The formula given is for one gallon, to simplify mixing. If you're about to smoke the meat from a whole pig, you'll need a lot more.

Pack the ham and bacon slabs loosely into a big crock. Cover with brine solution, and add a rock to keep them submerged. Set the crock in a cool spot (35° to 45°F.). The crock should be covered—it's very discouraging to find a pickled dead rat in it, particularly if you also find toothmarks on the hams.

The amount of time hams usually need to soak is one day per pound. That's for each individual ham, not by total poundage. In other words, if you have two fifteen-pound hams, soak fifteen days, not thirty. The minimum soaking period is two weeks, the maximum about one month. Move the hams around daily so the same areas are not always in contact with each other. Check the water to see that it doesn't turn sour, that is, milky and smelly. If it does, remove the hams at once, rinse well, and start over. But in that case, boil out the container before refilling.

When they have soaked sufficiently, rinse the hams and bacon off, dry well, and hang in the smokehouse. The smoldering fire for them should be started several hours before you tote them in, to ensure that temperature and smoke are adequate. Your smoky fire can be of hardwood chips—hickory for that hickory flavor, naturally—ground corncobs, or fruit-tree chips. Do not use birch, beech, or soft resinous woods unless you like strangely flavored hams.

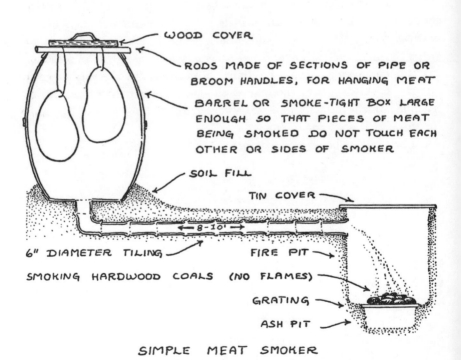

WOOD COVER

RODS MADE OF SECTIONS OF PIPE OR BROOM HANDLES, FOR HANGING MEAT

BARREL OR SMOKE-TIGHT BOX LARGE ENOUGH SO THAT PIECES OF MEAT BEING SMOKED DO NOT TOUCH EACH OTHER OR SIDES OF SMOKER

SOIL FILL

TIN COVER

← 8-10' →

6" DIAMETER TILING

FIRE PIT

SMOKING HARDWOOD COALS (NO FLAMES)

GRATING

ASH PIT

SIMPLE MEAT SMOKER

Smoke bacon continuously for one to two weeks, hams two to four weeks. After removing a ham from the smokehouse, cover it with a cheesecloth stocking or a coating of paraffin and hang in a cool place free from dampness and protected from flies and insects. An older method is to bury it in salt or in the grain bin, to exclude air. It will keep safely for three to five months in a cool, dry spot.

SAUSAGE MAKING

Sausages are to pork what hamburger is to beef. And they're just as simple to make. Take all your meat trimmings and cut off most of the fat, saving that for making lard. The meat trimmings include everything except the entrails that can't be used either as chops or roasts or cured. Feed them into a meat grinder. Add pepper, salt, and any herbs and spices you may like—bay leaves, caraway, cumin, sage, thyme, oregano—to the ground meat. The proportions are strictly up to individual taste, but, in my estimation, except for salt, it's pretty hard to overseason. Put the spice-and-meat mixture through the grinder again. What comes out is bulk sausage.

Fry it up like hamburger—but make sure it's well done. Pork in the United States is still often trichinosis-riddled. Also, don't let ground meat get warm sitting around between grindings. Always keep it cool or refrigerate it.

If you want to make real sausages, you'll need casings. Originally, casings were made from the small intestine of the pig that contributes the stuffing. For natural sausages, they're still the best. Make sure the pig doesn't eat for twenty-four hours before slaughtering and the amount of cleaning involved will be minimal. If you're not doing the slaughtering yourself, remind the butcher that you want the small intestine back for sausage-making.

The intestine is a long tube of skin. Turn it all inside out and wash thoroughly under running water. The smell won't be the greatest, since there will still be some half-digested food in it. After washing, turn the intestine, now a casing, right side out again. You're ready to stuff.

Synthetic cellulose casings can be bought and are considered edible. You've been eating them around hot dogs for years. If you don't want the work of cleaning intestines, though it's really not much work, you can settle for these. Usually they will need to be soaked in tepid water for fifteen minutes before use to soften them up.

Attach a funnel over the meat grinder where the meat comes out. Pull the casing over the tip, stocking-fashion, and you have an instant sausage-making machine. Hold the casing snugly around the tip of the funnel and fill as you grind the meat once more. Guide it with your hands as it winds its way across your kitchen work table, or it will break. Tie with string into links, any length you want. A two-foot sausage curled around a mound of scrambled eggs makes a fantastic family country breakfast.

Sausage will keep two or three weeks refrigerated and without preservatives. You can smoke-cure it, but all this does is improve the flavor; it doesn't enhance its keeping qualities. Experiment with other meat flavors. Try beef and pork sausages, or chicken and pork, for instance, and of course grind in a little garlic from your garden sometime, or wheat germ from your last trip to the miller's with a bag of grain.

THE LARD AND CRACKLINGS

Homemade lard has no resemblance at all to its plastic commercial counterpart. Would you believe real flavor?

Take the fat remaining after butchering a hog. This includes the fat from all trimmings, internal as well as external. Inspect it carefully for any chunks of meat. Cut them off. Chop the fat into one-inch squares.

Simmer the fat over low heat until it looks like well-done bacon fat. This usually takes about half an hour. Don't try to rush things. The heat must be kept low or you'll scorch your would-be lard.

When well done, scoop out the solids from the liquid fat. Place them on a wire screen over a pan in the oven. Set the oven at 200° to 225°F. This will drive out the last of the fat, which will collect in the drip pan. In about two hours the cracklings are dry and done. You'll probably eat half of them right then and there. Use what's left in meat gravies, or nibble on them like popcorn for a late evening snack. Cracklings will go stale in a week. If by any chance you have them around that long, the chickens will love to help you out.

The liquid lard should be strained through several layers of cheesecloth to remove any crackling fragments. Pour it into small stone crocks. Old coffee cans with plastic lids work fine if you don't have crocks around. Wait till the lard hardens before covering. If kept cool, lard can be stored for six months. In the refrigerator it will keep a year.

With your larder filled, let the winter gales blow; they'll bring nothing but contentment. Come spring, the root cellar will once more be empty, the kraut and hickory-smoked hams long since devoured, the dried fruit schnackled away in front of crackling winter fires. But the fields will be greening, the bees readying the hive for yet another harvest. The grand cycle of nature rolls on.

Catalog

*Knowledge is of two kinds: we know a subject
ourselves, or we know where we can find
information upon it.*
—SAMUEL JOHNSON

GOVERNMENT SERVICES

Government services indisputably lead the list of sources of farm information. Always start at the local level. Check the telephone book for your County Agent, County Extension Service, and Soil Conservation Service. They can take care of almost any problem you have, or refer you to someone who can.

You'll find another gold mine of farm problem solutions for your particular region in the State Agricultural Station. It publishes booklets and research findings concerning crops, livestock, and farm management for the state. Although understaffed, the station will help you with information if you have a specific problem and can explain it clearly enough. Also the publications make great reading. But stick with the agricultural experiment station in your own state; publications usually will not be sent to out-of-staters.

Alabama Agricultural Experiment Station, Auburn, Alabama.
Alaska Agricultural Experiment Station, College, Alaska.
Arizona Agricultural Experiment Station, Tucson, Arizona.

Arkansas Agricultural Experiment Station, Fayetteville, Arkansas.
California Agricultural Experiment Station, Berkeley, California.
Colorado Agricultural Experiment Station, Fort Collins, Colorado.
Connecticut Agricultural Experiment Station, New Haven, Connecticut.
Connecticut Agricultural Experiment Station, Storrs, Connecticut.
Delaware Agricultural Experiment Station, Newark, Delaware.
Florida Agricultural Experiment Station, Gainesville, Florida.
Georgia Agricultural Experiment Station, Experiment, Georgia.
Hawaii Agricultural Experiment Station, Honolulu, Hawaii.
Idaho Agricultural Experiment Station, Moscow, Idaho.
Illinois Agricultural Experiment Station, Urbana, Illinois.
Indiana Agricultural Experiment Station, Lafayette, Indiana.
Iowa Agricultural Experiment Station, Ames, Iowa.
Kansas Agricultural Experiment Station, Manhattan, Kansas.
Kentucky Agricultural Experiment Station, Lexington, Kentucky.
Louisiana Agricultural Experiment Station, University Station, Baton Rouge, Louisiana.
Maine Agricultural Experiment Station, Orono, Maine.
Maryland Agricultural Experiment Station, College Park, Maryland.
Massachusetts Agricultural Experiment Station, Amherst, Massachusetts.
Michigan Agricultural Experiment Station, East Lansing, Michigan.
Minnesota Agricultural Experiment Station, University Farm, St. Paul, Minnesota.
Mississippi Agricultural Experiment Station, State College, Mississippi.
Missouri Agricultural Experiment Station, Columbia, Missouri.
Montana Agricultural Experiment Station, Bozeman, Montana.
Nebraska Agricultural Experiment Station, Lincoln, Nebraska.
Nevada Agricultural Experiment Station, Reno, Nevada.
New Hampshire Agricultural Experiment Station, Durham, New Hampshire.
New Jersey Agricultural Experiment Station, New Brunswick, New Jersey.
New Mexico Agricultural Experiment Station, State College, New Mexico.

New York (Cornell) Agricultural Experiment Station, Ithaca, New York.

New York State Agricultural Experiment Station, Geneva, New York.

North Carolina Agricultural Experiment Station, State College Station, Raleigh, North Carolina.

North Dakota Agricultural Experiment Station, State College Station, Fargo, North Dakota.

Ohio Agricultural Experiment Station, Wooster, Ohio.

Oklahoma Agricultural Experiment Station, Stillwater, Oklahoma.

Oregon Agricultural Experiment Station, Corvallis, Oregon.

Pennsylvania Agricultural Experiment Station, University Park, Pennsylvania.

Puerto Rico Agricultural Experiment Station, Río Piedras, Puerto Rico.

Rhode Island Agricultural Experiment Station, Kingston, Rhode Island.

South Carolina Agricultural Experiment Station, Clemson, South Carolina.

South Dakota Agricultural Experiment Station, State College Station, South Dakota.

Tennessee Agricultural Experiment Station, Knoxville, Tennessee.

Texas Agricultural Experiment Station, College Station, Texas.

Utah Agricultural Experiment Station, Logan, Utah.

Vermont Agricultural Experiment Station, Burlington, Vermont.

Virginia Agricultural Experiment Station, Blacksburg, Virginia.

Virginia Truck Experiment Station, Norfolk, Virginia.

Washington Agricultural Experiment Station, Pullman, Washington.

West Virginia Agricultural Experiment Station, Morgantown, West Virginia.

Wisconsin Agricultural Experiment Station, Madison, Wisconsin.

Wyoming Agricultural Experiment Station, Laramie, Wyoming.

The largest source of published farm material is the federal government. Send for their "List of Available Publications of the United States Department of Agriculture" (45¢), Superintendent of Documents, U.S. Government Printing Office, Washington, D.C. 20402. Be prepared for two things, however: 50 percent of the booklets will be out of print, and if you send away for more than ten at a time, you probably won't even get a reply. Remember this is bureaucracy at its highest level.

EQUIPMENT

When it comes to ordering tools, farm implements and furnishings, and even baby chicks, the Sears and Ward catalogs are tried and true old standbys. Once you get your catalog, you have to keep reordering to receive the new one, so don't buy everything at once, even if you can afford it. Stagger your purchases over the year. This not only assures your getting the updated catalog but also delivers to the outhouse door a steady supply of john paper. Even before the recycling movement there was a Sears or Ward catalog hanging in every "half moon palace."

For more esoteric mail order fulfillment to even the most isolated RD address, try the following catalogs (free except where specified).

BEE KEEPING EQUIPMENT

Dadant & Sons Inc., Hamilton, Illinois 62341.
Kelley, Walter T., Co., Clarkson, Kentucky 42726.
Root, A. I., Co., Medina, Ohio 44256.

GOAT KEEPING EQUIPMENT

American Supply House, P.O. Box 1114, Columbia, Missouri 65201.
Breeders Supply Co., 101 South Main Street, Council Bluffs, Iowa 51501.
Dolly Enterprises, 279 Main, Colchester, Illinois 62326.
Hoegger Supply Co., Milford, Pennsylvania 18337.
Thiele, John H., P.O. Box 62, Warwick, New York 10990.
Tomellem Co., Calico Rock, Arkansas 72519.

HOUSEKEEPING

Cider Presses
Day Equipment Corp., 1402 East Monroe, Goshen, Indiana 46526.

Orchard Equipment and Supply Co., Conway, Massachusetts 01341.
Lamps, Kerosene
Aladdin Industries Inc., Kerosene Lamp Division, Nashville, Tennessee 37210.
Mills, Flour
Lee Engineering Co., 2023 West Wisconsin Avenue, Milwaukee, Wisconsin 53201.
Nelson & Sons, Inc., Box 1296, Salt Lake City, Utah 84110.
Mills, Grain
Smithfield Implement Co., 99 North Mark Street, Smithfield, Utah 84335.
Stoves, Wood Burning
Ashley Automatic Heater Co., P.O. Box 370, Sheffield, Alabama 35660.
Yard Goods and Country Clothes
Gohn Brothers, Middlebury, Indiana 46540.
Yogurt Culture
Daisyfresh Yogurt Co., P.O. Box 4295, Riverside, California 92504.

LAND MANAGEMENT
Chain Saws and Accessories
Homelite, Division of Textron, Port Chester, New York 10573.
McCulloch Corp., 6101 West Century Boulevard, Los Angeles, California 90045.
Granberg Industries, 201 Nevin, Richmond, California 94801.
Peat Starter Pots
Jiffy Products of America, P.O. Box 338, West Chicago, Illinois 60185.
Polyethylene Film
Dao Corp., Box 659, Terre Haute, Indiana.
Porta-green Co., 41 Fornof Road, Columbus, Ohio 43207.
Vite's, 2610 Redbud Trail, Niles, Michigan 49120.
Wayne Floral Co. Inc., P.O. Box 6, Newark, New York 14513.
Pruning Tools
Corona Clipper Co., Corona, California 91720.
Mellinger's, 2310 West South Range Road, North Lima, Ohio 44452.
Ratchet-Cut, Box 303, Milldale, Connecticut 06467.

Rain Gauges

Tru-Chek Rain Gauge Co., 100 North Broadway, Albert Lea, Minnesota 56007.

Seeders, Garden

Esmay Products Inc., Bristol, Indiana 46507.

Shredders, Compost

Amerind-MacKissic Inc., Box 111, Parker Ford, Pennsylvania 19457.

Gibson Bros. Co., P.O. Box 152, Plymouth, Wisconsin 53073.

Kemp Shredder Co., 627 Kemp Building, Erie, Pennsylvania 16512.

McCulloch Equipment Co., Box 3068, Torrance, California 90510.

Osborne Mfg. Co., P.O. Box 29, Osborne, Kansas 67473.

Red Cross Mfg. Corp., Bluffton, Indiana 46714.

Royer Foundry and Machine Co., 158 Pringle, Kingston, Pennsylvania 18704.

Winona Attrition Mill, 1009 West 5th Street, Winona, Minnesota 55987.

W-W Grinder Corp., Wichita, Kansas 67219.

Soil Testing Kits

Perfect Garden Co., 14 East 46th Street, New York, New York 10017.

Sudbury Laboratory, Sudbury, Massachusetts 01776.

Sprayers

Smith, D. B., & Co., 703 Main Street, Utica, New York 13501.

Tillers

Autohoe, West De Pere, Wisconsin 54178.

Geiger Corp., Box 2853, Harleysville, Pennsylvania 19438.

Gilson Brothers Co., P.O. Box 152, Plymouth, Wisconsin 53073.

Merry Manufacturing Co., P.O. Box 370, Edmonds, Washington 98020.

Roto-Hoe Co., Newbury, Ohio 44065.

Troy-Bilt, 102nd Street & Ninth Avenue, Troy, New York 12182.

Tractor Parts

Tractor Supply Co., 4747 North Ravenswood, Chicago, Illinois 60640.

Water Pumps

Baker Manufacturing Co., Evansville, Wisconsin 53536.

Dempster Industries, Beatrice, Nebraska 68310.

Windmills

Aermotor, 2500 West Roosevelt Road, Chicago, Illinois 60608.

Dempster Industries, Inc., Beatrice, Nebraska 68310

Dyna Technology, P.O. Box 3263, Sioux City, Iowa 51102.

Heller-Aller Co., Corner Perry and Oakwood, Napoleon, Ohio 43545.

TOOLS, MECHANICAL

Ben Meadows Co., P.O. Box 8377, Station F, 553 Amsterdam Avenue N.E., Atlanta, Georgia 30306.

Snap-on Tools Corp., Kenosha, Wisconsin 53140.

U.S. General Supply Corp., 100 General Place, Jericho, New York 11753 ($1.00).

TOOLS, WOODWORKING

Allcraft Tool & Supply Co., 215 Park Avenue, Hicksville, New York 11801 ($1.00).

Constantine, 2050 Eastchester Road, Bronx, New York 10461.

Silvo Hardware Co.. 107 Walnut Street, Philadelphia, Pennsylvania 19106 (25¢).

Woodcraft Supply Corp., 313 Montvale Avenue, Woburn, Massachusetts 01801 (25¢).

VETERINARY SUPPLIES

Eastern States Serum Co., 1727 Harden Street, Columbia, South Carolina 29204.

Kansas City Vaccine Co., Stock Yards, Kansas City, Missouri 64102.

United Pharmical Co., 8366 La Mesa Boulevard, La Mesa, California 92041.

FERTILIZERS

Most of the following firms not only will ship organic fertilizers and soil developers, some of them even do it by the railroad carful.

There are probably other nurseries and supply stores in your region handling natural fertilizers. Send away for some extra price lists anyhow; comparison shopping by mail can often save lots of money.

ARKANSAS

Naturail Resources Development, Grannis, Arkansas 71944.

CALIFORNIA

California Eweson Corp., 2840 Hidden Valley Lane, Santa Barbara, California 93103.

Ecology Trading Center, 788 Old Country Road, Belmont, California 94002.

Eden Acres Farm, 1264 East Alvarado Street, Fallbrook, California 92028.

Fersolin Corp., 100 Bush Street, San Francisco, California 94104.

Forci-Grow, Route 1, Box 1866, Lathrop, California 95330.

Ocean Labs, Inc., Berth 42 Outer Harbor, San Pedro, California 90731.

Vita Green Farms, P.O. Box 878, Vista, California 92083.

Wright Feeds, 16210 South Colorado, Paramount, California 90723.

COLORADO

Rich Loam, Colorado Springs, Colorado 80900.

U Need Us Fertilizer Co., Austin, Colorado 81410.

CONNECTICUT

Biological Control Laboratories, Box 867, Groton, Connecticut 06340.

Brookside Nurseries Inc., 228 Brookside Road, Darien, Connecticut 06820.

Pine Willow Farms, North Taylor Avenue, South Norwalk, Connecticut 06854.

Sea-Born, Mineral Division of Skod Co., 10 Lewis Street, Greenwich, Connecticut 06830.

FLORIDA
Atlantic & Pacific Research Inc., Box 14366, North Palm Beach, Florida 33403.

Conklin, James B., 701 West Highway 54, Zephyrhills, Florida 33599.

Lee's Fruit Co., Box 450, Leesburg, Florida 32748.

Neier, Carlton W., 6730 South Drive, Melbourne, Florida 32901.

GEORGIA
Blenders Inc., Lithonia, Georgia 30058.

ILLINOIS
Home and Garden Supply Co., 4701 West 55th Street, Chicago, Illinois 60632.

Klueter Feed Store Inc., Route 159, Edwardsville, Illinois 62025.

Lake-Cook Farm Supply Co., 997 Lee Street, Des Plaines, Illinois 60016.

Oil-Dri Corp. of America, 520 North Michigan Avenue, Chicago, Illinois 60611.

Sea-Born, Mineral Division of Skod Co., 3421 North Central Avenue, Chicago, Illinois 60634.

Soil Life Research Co., Box 132, Paw Paw, Illinois 61353.

INDIANA
Associated Specialists, P.O. Box 21314, Indianapolis, Indiana 46221.

Norwegian Sea Kelp Prods., 2223 Lafayette Road, Indianapolis, Indiana 46222.

IOWA
Brisbois, F. S., Fonda, Iowa 50540.

Hy-Brid Sales Co., P.O. Box 276, Council Bluffs, Iowa 51501.

Norwegian Sea-Weed Products, 206 Avenue D, Ft. Madison, Iowa 52627.

KANSAS

Wonder Life Co., 322 South Summit, Arkansas City, Kansas 67005.

KENTUCKY

Happy Acres Inc., P.O. Box 711, Somerset, Kentucky 42501.
Stille, Clair W., 130 North Hanover Avenue, Lexington, Kentucky 40502.

MARYLAND

Yoder, Eli, Route 2, Box 129, Oakland, Maryland 21550.

MASSACHUSETTS

Alexander, J. Herbert, Dahliatown Nurseries, Middletown, Massachusetts 02346.
Shady Lane Greenhouse, P.O. Box 43, East Princeton, Massachusetts 01517.
Squanto Peat & Organic Fertilizer Co., Oakham, Massachusetts 01068.

MICHIGAN

Allen's Lawn & Garden Store, 2925 Francis Street, Jackson, Michigan 49203.
Fanning Soil Service Inc., 4951 South Custer Road, Monroe, Michigan 48161.
Graszow's Organic Garden, 8520 Dixie Highway, Clarkston, Michigan 48016.
Juengel, Theo. T., 11 East Grove Street, Sebawaing, Michigan 48759.
Uncle Luke's Feed Store, 6691 Livernois, Troy, Michigan 48084.

MINNESOTA

Perkins Crosslake Garden Center, Crosslake, Minnesota 56442.

NEW JERSEY

Far Hills Nursery Inc., Bedminster, New Jersey 07921.

Hyper-Humus Co., P.O. Box 267, Newton, New Jersey 07860.

NEW MEXICO

New Mexico Humates, Inc., 701 Madison N.E., Albuquerque, New Mexico 87110.

NEW YORK

Almith Industries, Box 275, Seaford, New York 11783.

Hindson, RD 1, Honeoye Falls, New York 14472.

Massapequa Seed & Garden Supply, 500–504 Hicksville Road, Massapequa, New York 11758.

Sterling Forest Peat Co., Box 608, Tuxedo, New York 10987.

OHIO

Nature's Way Products, 3505 Mozart Avenue, Cincinnati, Ohio 45211.

Windy Hill, 6046 Benken Lane, Cincinnati, Ohio 45211.

OKLAHOMA

Delta Mining Corp., P.O. Box 95, Mill Creek, Oklahoma 74856.

Natural Plant Food Co., 1409 N.W. 50th Street, Oklahoma City, Oklahoma 73118.

PENNSYLVANIA

Best Feeds & Farm Supplies Inc., Oakdale, Pennsylvania 15071.

Degler, Paul, 51 Bethlehem Pike/Route 309, Colmar, Pennsylvania 18915.

Leap, Don, Hyndman, Pennsylvania 15545.

Natural Development Co., Box 215, Bainbridge, Pennsylvania 17502.

Zook & Ranck Inc., RD 1, Gap, Pennsylvania 17527.

TENNESSEE

Jones, Robin, Phosphate Co., 204 23rd Avenue, Nashville, Tennessee 37206.

TEXAS

Garden Mart, 5108 Bissonnet Street, Bellaire, Texas 77401.
Minores Inc., 4614 Sinclair Road, San Antonio, Texas 78222.

UTAH

Key Minerals Corp., P.O. Box 2364, Salt Lake City, Utah 84110.

VERMONT

Davidson, Donald G., South Royalton, Windsor County, Vermont 05001.

VIRGINIA

Greenlife Products Co., West Point, Virginia 23181.
Standard Products Co. Inc., Kilmarnock, Virginia 22482.
Tangier Sea Organism Co., Sinclairs Beach, Tangier, Virginia 23440.

WASHINGTON

Alaska Fish Fertilizer, 84 Seneca Street, Seattle, Washington 98101.
Evergreen Organic Supplies, Route 2, Box 839, Sultan, Washington 98294.
Organic Gardens, Route 1, Box 31, Soap Lake, Washington 98851.

WISCONSIN

Grove Compost Co. Inc., P.O. Box 242, Grafton, Wisconsin 53024.
Kirchner, Route 2, Shiocton, Wisconsin 54170.
Lee, Mosser, Co., Millston, Wisconsin 54643.
Slug, W. G., Seed & Fertilizer Inc., 3904–3922 West Villard Avenue, Milwaukee, Wisconsin 53209.

CANADA

Annapolis Valley Peat Moss Co. Ltd., Berwick, Nova Scotia, Canada.

Bedford Organic Fertilizer Co., 2045 Bishop Street, Montreal, P.Q., Canada.

Berman, Ted, RR 2, Leamington, Ontario, Canada.

MacDonald & Wilson, 562 Beatty Street, Vancouver, B.C., Canada.

LIVESTOCK

In many cases you will be able to buy your livestock locally. And that is, in fact, recommended. However, should you want some baby chicks right away, the best thing to do is order them by mail. In some places this is now almost the only way for farmers to get chicks, since hatching them has become such a specialized operation.

Pigs should present no problem. As soon as you get to know more than two people, at least one will either have some pigs he'll sell you or will know where they are available. Goats, on the other hand, are far less common. The easiest way to locate some of the breed that appeals to you is to write to the breed's national headquarters or the American Dairy Goat Association and inquire about the farm nearest you.

Insects are not usually considered livestock, but bees are some of the best workers you can have around. And they do move—which is why they're in this section, along with other beneficial insects such as praying mantises and ladybugs.

BEES

Alamance Bee Co., La Belle, Florida 33935.

Bessonet Bee Co. Inc., Donaldsonville, Louisiana 70346.

Calvert Apiaries, Calvert, Alabama 36513.

Crenshaw County Apiaries, Rutledge, Alabama 36071.

Harper Apiaries, New Brockton, Alabama 36351.

Jensen's Apiaries, Macon, Mississippi 39341.

Kane Apiaries, Hallettsville, Texas 77964.

Leverette, Walter D., Apiaries, P.O. Box 3364, Ft. Pierce, Florida 33450.

McCary, G.D., & Son Bee Co., P.O. Box 87, Buckatunna, Mississippi 39322.

Merrill Bee Co., Box 2687, State Line, Mississippi 39362.

Mitchell's Apiaries, Bunkie, Louisiana 71322.

Norman Bee Co., Route 1, Ramer, Alabama 36069.

Spalding, Michael, Box 133, Ringoes, New Jersey 08851.

Stover Apiaries Inc., Mayhew, Mississippi 39753.

Stricker, M. H., Stockton, New Jersey 08559.

Trent Valley Apiaries, Route 2, Box 193, Trent, North Carolina 08559

Weaver, Howard, & Sons, Navasota, Texas 77868.

Wilbanks Apiaries, Claxton, Georgia 30417.

York Bee Company, P.O. Box 307, Jesup, Georgia 31545.

FOWL

Grain Belt Hatchery, Windsor, Missouri 65360.

Hockmans, Box 718724, San Diego, California 92107.

Hoffman Hatchery, Gratz, Pennsylvania 17030.

Marti Poultry Farm, Windsor, Missouri 65360.

Murray McMurray Hatchery, Webster City, Iowa 50595.

Stromberg's, Fort Dodge, Iowa 50501.

Surplus Chick Co., Milesburg, Pennsylvania 16853.

Vernon Norris Hatchery, Mars-Cooperstown Road, Valencia, Pennsylvania 16059.

GOATS (CLUBS)

Alpines International, Mrs. Dorothy Burke, President, Route 4, Box 315, Tarentum, Pennsylvania 15084.

American Dairy Goat Association, P.O. Box 186, Spindale, North Carolina 28160.

American La Mancha Club, Mrs. Carl Erbe, President, J15616 East Santos Avenue, Ripon, California 95366.

National Nubian Club, George Randle, President, Placitas, New Mexico 87043.

National Toggenburg Club, Mrs. Elizabeth Gustafson, President, Route 1, Box 103, Grandview, Washington 98930.

INSECTS, BENEFICIAL

Eastern Biological Control, County Line Road, Jackson, New Jersey 08527.

Ecology Trading Center, 788 Old Country Road Belmont, California 94002.

Bank, Ted P., 608 Eleventh Street, Pitcairn, Pennsylvania 15140.

Bio-Control Co. (the old California Bug Co.), Route 2, Box 2397, Auburn, California 95603.

Fairfax Biological Laboratories, Clinton Corners, New York 12514.

Lakeland Nurseries Sales, Insect Control Division, Hanover, Pennsylvania 17331.

Gothard Inc., P.O. Box 370, Canutillo, Texas 79835.

Mince-Moyers Nursery, County Line Road, Jackson, New Jersey 08527.

Schnoor, L.E., Rough & Ready, California 95975.

PLANTS

When ordering from seed and nursery companies, remember their illustrators think they're a cross between Walt Disney and Crumb, not to mention that the sales pitch is often slipperier than a greased pig. Take what you read in the catalogs with a grain of salt. Your plants will grow and thrive, but don't be disappointed if not all of them are picture-perfect. Speaking of pictures, don't fail to order a Shumway catalog—the old steel engravings are fabulous.

FRUITS AND VEGETABLES

Alexander, J. Herbert, Dahliatown Nurseries, Middleboro, Massachusetts 02346.

Burpee, W. Atlee, Co., Philadelphia, Pennsylvania 19132. Ask for Blue List if you intend to grow any vegetables in quantity.

Breck's, 200 Breck Building, Boston, Massachusetts 02210.

Burgess Seed and Plant Co., Galesburg, Michigan 49053.

Burrell Seed Co., Box 150, Rocky Ford, Colorado 81067.

Foster, Dean, and Sons Nursery, Hartford, Michigan 49057.

Gurney Seed and Nursery Co., Yankton, South Dakota 57078.

Harris, Joseph, Co. Inc., Moreton Farm, Rochester, New York 14624.

Jung, J. W., Seed Co., Randolph, Wisconsin 53956.

Livingston Seed Co., P.O. Box 299, Columbus, Ohio 43216.

Lowdon, Edward, Ancaster, Ontario, Canada.

Makielski Berry Farm, 7130 Platt Road, Ypsilanti, Michigan 48197.

Natural Development Co., Bainbridge, Pennsylvania 17502.

Park, Geo. W., Seed Co. Inc., P.O. Box 31, Greenwood, South Carolina 29646.

Rayner Brothers Inc., Salisbury, Maryland 21801.

Robson Quality Seeds Inc., Hall, New York 14463.

Savage Farm Nursery, P.O. Box 125, McMinnville, Tennessee 37110.

Shumway, R. H., Seedsman, P.O. Box 777, Rockford, Illinois 61101.

Spring Hill Nurseries, Tipp City, Ohio 45371.

Vita Green Farms, P.O. Box 878, Vista, California 92083. Large selection of untreated, unhybridized seeds.

TREES

Carino Nurseries, Box 538, Indiana, Pennsylvania 15701.

Chickadee Nursery, Sherwood, Oregon 97140.

Gerardi, Louis, Nursery, Route 1, O'Fallon, Illinois 62269.

Girard Nurseries, Geneva, Ohio 44041.

Millcreek Associates, Box 178, Warrington, Pennsylvania 18976.

Miller, J. E., Nurseries, Canandaigua, New York 14424.

Musser Forests Inc., Indiana, Pennsylvania 15701.

Ozark Nurseries Garden Center, Route 2, Tahlequah, Oklahoma 74464.

Stark Brothers Nurseries, Louisiana, Missouri 63353.

Tennessee Nursery Co., Box 4, Cleveland, Tennessee 37311.

Waynesboro Nurseries, Waynesboro, Virginia 22980.

Western Maine Forest Nursery Co., Fryeburg, Maine 04037.

PUBLICATIONS

Farm magazines are a valuable source of practical ideas, and their advertisements give you a feel for not only what is available, but costs in general. You'll have to be selective, of course, about the information used—by far the greatest number of publications are oriented toward chemical pesticides and fertilizers.

There is obviously no reason to subscribe to even a tenth of the periodicals listed here. Pick and choose those that are either of regional importance or particularly appropriate to your interests. Then send away for a sample copy to see if a subscription would be really helpful. Don't be surprised, by the way, if you happen to get a reply telling you you can't subscribe. *Farm Journal,* for instance, rarely sends subscriptions to persons not having a RD address, and some of the state publications do not accept out-of-state subscribers.

GENERAL

Agricultural Education Magazine, Box 5115, Madison, Wisconsin 53705.

Alabama Farmer, Southern Farm Publications, Box 6429, Nashville, Tennessee 37212.

American Agriculturist & The Rural New Yorker, Savings Bank Building, Ithaca, New York 14850.

American Farm Youth, Youth Publishing Co., 113 West Main, Danville, Illinois 61833.

Arizona Farmer-Ranchman, 434 West Washington Street, Phoenix, Arizona 85003.

Arkansas Farmer, Southern Farm Publications, Box 6429, Nashville, Tennessee 37212.

Big Farmer, 534 North Broadway, Milwaukee, Wisconsin 53202.

Bio-Dynamics, RD 1, Stroudsburg, Pennsylvania 18360.

Buckeye Farm News, 245 North High Street, Columbus, Ohio 43216.

California Farmer, 83 Stevenson Street, San Francisco, California 94105.

California Grange News, 2101 Stockton Boulevard, Sacramento, California 95817.

California Rancher, Sierra Printing & Publishing Co., 2900 Rio Linda Boulevard, Sacramento, California 95815.

Dakota Farmer, 1216 South Main Street, Aberdeen, South Dakota 57401.

Electricity on the Farm Magazine, R. H. Donnelley Corp., 466 Lexington Avenue, New York, New York 10017.

Empire State Grange, Box 368, Ithaca, New York 14850.

Farm Journal, 230 West Washington Square, Philadelphia, Pennsylvania 19106.

Farm Quarterly, 22 East 12th Street, Cincinnati, Ohio 45210.

Farmer-Stockman, Box 25125, Oklahoma City, Oklahoma 73125.

Florida Agriculture, Box 7605, Orlando, Florida 32804.

Free Press Weekly, 300 Carlton Street, Winnipeg 2, Manitoba, Canada.

Georgia Farmer, Southern Farm Publications, Box 6429, Nashville, Tennessee 37212.

Good Farming, 1450 Don Mills Road, Don Mills, Ontario, Canada.

High Plains Journal, Box 760, 1500 East Wyatt Earp Boulevard, Dodge City, Kansas 67801.

Hoosier Farmer, Indiana Farm Bureau Inc., 130 East Washington Street, Indianapolis, Indiana 46204.

Idaho Agricultural Science, College of Agriculture, University of Idaho, Moscow, Idaho 83843.

Iowa Farm Science, Iowa State University of Science & Technology, Ames, Iowa 50010.

Kansas Farmer, Harvest Publishing Co., 719 Mills Bldg., Topeka, Kansas 66612.

Kentucky Farmer, Southern Farm Publications, Box 6429, Nashville, Tennessee 37212.

Louisiana Farmer, Southern Farm Publications, Box 6429, Nashville, Tennessee 37212.

Mississippi Farmer, Southern Farm Publications, Box 6429, Nashville, Tennessee 37212.

National Future Farmer, Future Farmers of America, Alexandria, Virginia 22306.

New Mexico Farm & Ranch, 421 North Water Street, Las Cruces, New Mexico 88001.

North Carolina Grower, Southern Farm Publications, Box 6429, Nashville, Tennessee 37212.

Ohio Farmer, Harvest Publishing Co., 1350 West Fifth Avenue, Columbus, Ohio 43212.

Oklahoma Ranch & Farm World, Tulsa World, 315 South Boulder, Tulsa, Oklahoma 74102.

Organic Gardening & Farming, 33 East Minor Street, Emmaus, Pennsylvania 18049.

Prairie Farmer, 1230 Washington Boulevard, Chicago, Illinois 60607.

Progressive Farmer, 821 North 19th Street, Birmingham, Alabama 35203.

Redwood Rancher, Box 1, St. Helena, California 94574.

South Carolina Farmer-Grower, Southern Farm Publications, Box 6429, Nashville, Tennessee 37212.

South Dakota Farm & Home Research, South Dakota State University, Agricultural Experiment Station, Brookings, South Dakota 57006.

Successful Farming, Meredith Corp., 1716 Locust Street, Des Moines, Iowa 50303.

Tennessee Farm & Home Science, Agricultural Experiment Station, Knoxville, Tennessee 37916.

Tennessee Farmer, Southern Farm Publications, Box 6429, Nashville, Tennessee 37212.

Texas Agriculture, Texas Farm Bureau, Box 489, 401 Franklin Avenue, Waco, Texas 76703.

Western Crops & Farm Management, Nelson R. Crow Publications Inc., 251 Kearny Street, San Francisco, California 94108.

LIVESTOCK

BEES

American Bee Journal, Hamilton, Illinois 62341.

Canadian Bee Journal, 30 Victoria Street North, Port Hope, Ontario, Canada.

Gleanings in Bee Culture, Medina, Ohio 44256.

FOWL

American Poultry & Hatchery News, 521 East 63rd Street, Kansas City, Missouri 64110.

Canadian Poultry Review, Donovan Ltd., 129 Adelaide Street West, Toronto 1, Ontario, Canada.

Poultry, Queens Printer, Montreal, Quebec, Canada.

Poultry Digest, Garden State Bldg., Sea Isle City, New Jersey 08243.

Poultry Press, Box 947, York, Pennsylvania 17405.

Poultry Tribune, Watt Publishing Co., Mount Morris, Illinois 61054.

Virginia Poultryman, Virginia State Poultry Federation, 1001 East Main, Richmond, Virginia 23219.

GOATS

Dairy Goat Guide, 318 Waterloo Road, Marshall, Wisconsin 53559.

Dairy Goat Journal, Box 836, Columbia, Missouri 65201.

Sheep & Goat Raiser, Box 1840, San Angelo, Texas 76901.

PIGS

American Landrace, Culver Press, Culver, Indiana 46511.

Chester White Journal, Box 228, Rochester, Indiana 46975.

Duroc News, 239 N.E. Monroe, Peoria, Illinois 61602.

Hampshire Herdsman, 1111 Main Street, Peoria, Illinois 61606.

Hog Farm Management, Miller Publishing Co., 2501 Wayzata Boulevard, Minneapolis, Minnesota 55440.

Hog Production, Box 418, Richmond Hill, Ontario, Canada.

National Hog Farmer, 1999 Shepard Road, St. Paul, Minnesota 55102.

Poland China World, 501 East Losey Street, Galesburg, Illinois 61401.

Spotted News, Spotted Swine Record Inc., Bainbridge, Indiana 46105.

FISH POND

American Fishes & U.S. Trout News, 67 W 9000 S, Sandy, Utah 84070.

Farm Pond Harvest, Box 884, Kankakee, Illinois 60901.

WOODLANDS

American Forests, American Forestry Assn., 919 17th St. N.W., Washington, D.C. 20006.

Forest Notes, 5 South State Street, Concord, New Hampshire 03301.

Forest Science, Society of American Foresters, 1010 16th St. N.W., Washington, D.C. 20036.

Ohio Woodlands, The Neil House, Columbus, Ohio 43215.

Pennsylvania Forests, 5221 East Simpson Street, Mechanicsburg, Pennsylvania 17055.

READING FURTHER

Since natural farming is still very much in its infancy, source material for further reading is limited. As a matter of fact, outside of the

Rodale organic books there really isn't much. This goes for livestock as well as crops. Most of today's livestock not only eat pesticide-polluted feed, but are given questionable hormone injections for more rapid growth as well.

However, the following list of books should provide interesting reading for the apprentice farmer in search of more detailed information in specific areas. Just remember you'll have to dig between the chemical-covered lines to find what can be used on a natural homestead.

CROPS

Ahlgren, G. H. *Forage Crops*. New York: McGraw-Hill, 1956.

Brickbauer, E. A., and W. P. Mortenson. *Approved Practices in Crop Production*. Danville, Ill.: Interstate, 1967.

Dickson, J. G. *Diseases of Field Crops*. New York: McGraw-Hill, 1956.

Faulkner, E. H. *Plowman's Folly*. Norman, Okla.: University of Oklahoma Press, 1943.

Faulkner, E. H. *Soil Development*. Norman, Okla.: University of Oklahoma Press, 1957.

Guise, C. H. *Management of Farm Woodlands*. New York: McGraw-Hill, 1950.

Hughes, H. D., and E. R. Henson. *Crop Production*. New York: Macmillan, 1957.

Kipps, M. S. *Production of Field Crops*. New York: McGraw-Hill, 1970.

Mitchell, R. L. *Crop Growth and Culture*. Ames, Iowa: Iowa State University Press, 1970.

Rodale, J. I., *et al. The Encyclopedia of Organic Gardening*. Emmaus, Pa.: Rodale Books, 1970.

Rodale, J. I., *et al. How to Grow Vegetables and Fruits by the Organic Method*. Emmaus, Pa.: Rodale Books, 1970.

Shirley, H. L., and P. F. Graves. *Forest Ownership for Pleasure and Profit*. Syracuse, N.Y.: Syracuse University Press, 1967.

Tukey, H. B. *Dwarfed Fruit Trees*. New York: Macmillan, 1964.

Tyler, H. *Organic Gardening Without Poisons*. New York: Van Nostrand-Reinhold, 1970.

Wilson, H. K. *Grain Crops*. New York: McGraw-Hill, 1955.

EQUIPMENT AND MACHINERY

Bainer, R., *et al. Principles of Farm Machinery*. New York: Wiley, 1955.

Boshoff, W. H. *Using Field Machinery*. New York: Oxford University Press, 1968.

Hartzog, D. H. *Shop Guide in Farm Mechanics*. Danville, Ill.: Interstate, 1956.

Hunt, D. *Farm Power and Machinery Management*. Ames, Iowa: Iowa State University Press, 1968.

Nichols, B. *Machinery Around Your Country Home*. New York: Devin, 1969.

Smith, H. P. *Farm Machinery and Equipment*. New York: McGraw-Hill, 1964.

DISEASES (PLANT AND LIVESTOCK)

Anderson, H. W. *Diseases of Fruit Crops*. New York: McGraw-Hill, 1956.

Barron, N. S., and R. P. Feltwell. *Poultry Farmers Veterinary Book*. Springfield, Ill.: C. C. Thomas, 1959.

Dickson, J. G. *Diseases of Field Crops*. New York: McGraw-Hill, 1956.

Jones, F. G., and M. G. Jones. *Pests of Field Crops*. New York: St. Martin, 1964.

Kirk, R. W., and S. I. Bistner. *Handbook of Veterinary Procedures and Emergency Treatment*. Philadelphia: Saunders, 1969.

Metcalf, C. L. *Destructive and Useful Insects*. New York: McGraw-Hill, 1962.

Monnig, H. O., and F. S. Veldman. *Handbook of Stock Diseases*. San Francisco: Tri-Ocean, 1966.

Shurtleff, M. C. *How to Control Plant Diseases in Home and Garden*. Ames, Iowa: Iowa State University Press, 1966.

Stamm, G. W. *Veterinary Guide for Farmers*. New York: Hawthorn, 1963.

Whitney, L. F. *Farm Veterinarian*. Springfield, Ill.: C. C. Thomas, 1964.

GENERAL

Agriculture Staff of Iowa State University. *Midwest Farm Handbook*. Ames, Iowa: Iowa State University Press, 1969.

Ashby, W., *et al. Modern Farm Buildings*. Englewood, N.J.: Prentice-Hall, 1959.

Bromfield, L. *Malabar Farm*. New York: Ballantine, 1970.

Bullock, B. F. *Practical Farming for the South*. Chapel Hill, N.C.: University of North Carolina Press, 1946.

Efferson, J. N. *Principles of Farm Management*. New York: McGraw-Hill, 1953.

Evans, E. F., and R. L. Donahue. *Exploring Agriculture*. Englewood, N.J.: Prentice-Hall, 1963.

Hamilton, J., and W. R. Bryant. *Profitable Farm Management*. Englewood, N.J.: Prentice-Hall, 1963.

James, S. C. *Midwest Farm Planning Manual*. Ames, Iowa: Iowa State University Press, 1968.

Juergenson, E. M. *Farming Programs For Small Acreages*. Danville, Ill.: Interstate, 1959.

McMillen, W., ed. *Harvest: An Anthology of Farm Writing*. New York: Appleton-Century, 1964.

Pearson, L. C. *Principles of Agronomy*. New York: Van Nostrand-Reinhold, 1967.

Sloane, E. *Age of Barns*. New York: Funk & Wagnalls, 1966.

Vogt, E. F. *Modern Homesteaders: The Life of a Twentieth-Century Frontier Community*. Cambridge, Mass.: Harvard University Press, 1955.

LIVESTOCK

Acker, D. C. *Animal Science and Industry*. Englewood, N.J.: Prentice-Hall, 1962.

Anderson, A. L., and J. J. Kiser. *Introducing Animal Science*. New York: Macmillan, 1963.

Ashbrook, F. G. *Butchering, Processing and Preservation of Meat*. New York: Van Nostrand-Reinhold, 1955.

Baker, J. P., and T. B. Keith. *Feed Formulation Manual*. Danville, Ill.: Interstate, 1967.

Biddle, G. H., and E. M. Juergenson. *Approved Practices in Poultry Production.* Danville, Ill.: Interstate, 1963.

Brakensick, E. L., and L. J. Phipps. *Self-Study Guide in Animal Science.* Danville, Ill.: Interstate, 1966.

Byerly, T. C. *Livestock and Livestock Products.* Englewood, N.J.: Prentice-Hall, 1963.

Carroll, W. E., *et al. Swine Production.* New York: McGraw-Hill, 1962.

Cole, H. H. *Introduction to Livestock Production.* San Francisco: W. H. Freeman, 1966.

Cunha, T. J. *Swine Feeding and Nutrition.* New York: Wiley, 1957.

Eckert, J. E., and F. R. Shaw. *Beekeeping.* New York: Macmillan, 1960.

Ensminger, M. E. *Swine Science.* Danville, Ill.: Interstate, 1961.

Frish, K. Von. *Dancing Bees.* New York: Harcourt Brace, 1965.

Hammond, J. *Farm Animals.* New York: St. Martin, 1970.

Leach, C. A., and C. E. Leach. *Aids to Goatkeeping.* Columbia, Missouri: *Dairy Goat Journal,* 1961.

Mackenzie, D. *Goat Husbandry.* New York: Transatlantic, 1968.

Nordby, J. E., and H. E. Lattig. *Selecting, Fitting and Showing Poultry.* Danville, Ill.: Interstate, 1964.

Peck, W. D. *Pig Keeping.* New York: Transatlantic, 1960.

Ribbands, C. R. *Behavior and Social Life of Honey Bees.* New York: Dover, 1953.

Stewart, M. W. *Ducks, Turkeys and Geese.* San Francisco: Tri-Oceans, 1960.

---

*Whoever acquires knowledge and does not
practice it resembles him who ploughs his
land and leaves it unsown.*
—SA'DI

Index